MAUSER
MILITARY RIFLE MARKINGS

2nd Edition

Terence W. Lapin

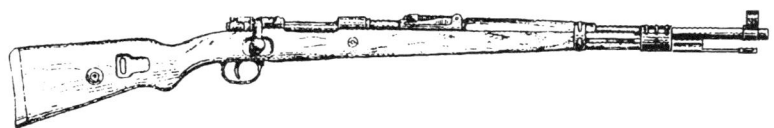

HYRAX PUBLISHERS, LLC

Copyright ©2001 by Terence W. Lapin. All rights reserved.
No part of this publication may be reproduced, translated, stored in a retrieval system, or transmitted in any form or by any means (electronic, mechanical, photocopy, recording, or otherwise) without the express prior written consent of the copyright holder.

ISBN 0-9676896-3-5

Hyrax Publishers, LLC
P.O. Box 10885
Arlington, VA 22210
www.hyraxbooks.com

Printed in the United States of America

Also by Terence W. Lapin

The Mosin-Nagant Rifle

The Soviet Mosin-Nagant Manual

The Soviet Tokarev Rifle Service Manual

Mannlicher Model 95 Rifle and Carbine:
The Royal Italian Infantry Manual

MAUSER MILITARY RIFLE MARKINGS

TABLE OF CONTENTS

Page

i	**Table of Contents**
ii	**Introduction**
vi	**Organization**
1	**Languages**
10	**Alphabets**
16	**Numbers**
21	**Calendars and Dating**
27	**Inscriptions in Western Languages**
81	**Inscriptions in Non-Western Languages**
94	**Coats of Arms**
114	**Emblems**
144	**German Wartime Codes**
154	**German World War II Subcontractor Codes**
157	**German Police District Codes**
161	**Gazetteer**
166	**Bibliography**

MAUSER MILITARY RIFLE MARKINGS

I can tell at sight a Mauser rifle from a javelin
--- Major-General Stanley
The Pirates of Penzance

INTRODUCTION

Most of us, like Gilbert and Sullivan's "very model of a modern Major-General", can distinguish a rifle from a long, pointed stick; being able to say anything more definite about a particular type of Mauser rifle is another matter entirely.*

This book is neither a history of Mauser rifles nor a catalogue of their many types and variations; readers interested in these aspects of the subject are best advised to consult the estimable works of Ludwig Olson, Robert Ball, Joe Poyer and Steve Kehaya, and others. This book is, rather, an attempt to compile in one place a general guide to reading and understanding the often-mystifying markings found on military Mauser rifles. I did not include hunting, target, and other non-military rifles because I preferred to keep the subject within manageable bounds.

Because Mauser rifles can sometimes be attributed to a specific country or entity only by a national shield or similar insignia stamped somewhere on the rifle I have included a chapter giving the "coats of arms" of countries and organizations known or believed to have used or made these rifles. I refrained from giving the locations of markings on specific rifles for two reasons: first, markings

---

* Readers of a musical bent may be interested to know that the original 1880 lyrics were "*chassepôt rifle*"; these were changed in later performances to "*Mauser rifle*" as *Mauser* became increasingly more familiar than *chassepôt* to English audiences. It is also easier to sing.

MAUSER MILITARY RIFLE MARKINGS

can and do appear in several places on various rifle models; and second, anyone looking up a particular marking will probably have found it on the weapon already and, consequently, will not need to be told where it is. If the location is significant for some reason, I have mentioned it.

Other sections of this book contain illustrations of inscriptions in foreign languages ---both in Roman letters and foreign scripts--- found on the rifles, which I have translated into English. I have included tables of several alphabets in which some of these inscriptions are written in the hope that readers will be interested in what the individual letters look like and the sounds they represent. It is also possible that readers may have some object, quite unrelated to firearms, on which the letters appear and whose place of origin they can identify by using the tables. In a similar vein I have included tables of foreign numerals; these can be used to decipher serial numbers, model numbers, dates, and so forth.

I did not believe that merely listing the markings was sufficient: in many instances that would be only minimally informative and not overly helpful. Some readers, I felt, really would like to know, for example, how and why **ABL** represents four words in two different languages, what the Orange Free State was, and other ---often admittedly esoteric--- bits of Mauser lore, such as Haile Selassie's real name and why a South American country (Uruguay) is an Oriental republic. For that reason I added explanatory passages titled **Notes** at the end of several chapters.

Anyone inspecting some of the more exotic Mausers will, inevitably, come across strange-looking sets of numbers which, because of their placement, do not appear to be (and often are not) serial numbers. Sometimes these sets are numbers identifying a specific model, but they can just as easily be the date of manufacture. In the

MAUSER MILITARY RIFLE MARKINGS

latter instance they will usually be completely incomprehensible to the reader as they are based on a local calendar. To this end I have furnished a brief description of the several dating systems used on Mausers, and have provided guides for converting such dates to A.D.

    A subject of major interest to many Mauser enthusiasts is the codes used by the Germans before and during the Second World War either to hide the source of a weapon's production or merely to indicate the acceptance of an individual weapon into service. These codes are available from any number of sources, but I included them in this volume for the sake of having such information in one place with the rest of the materials; once more, this is simply a convenience for the reader. I did not include individual regimental markings; those can be found in specialized volumes on the matter.

    I do not claim that this book is all-inclusive or definitive. I have certainly omitted a number of markings, and I crave the reader's indulgence for this. Nonetheless, I hope this book will provide anyone interested in Mauser military rifles with some useful information, some food for thought, or, failing all else, some idle entertainment.

Terence W. Lapin
Arlington, VA 2000

MAUSER MILITARY RIFLE MARKINGS

INTRODUCTION TO THE SECOND EDITION

The first edition of this book sold out more quickly than anticipated; in the interval between its publication and this one I acquired enough additional Mauser lore to justify a new edition, rather than a second printing of the first one.

Besides furnishing the various new markings listed and described in this edition I have overcome my reluctance, referred to in the Introduction to the first edition, to go into the subject of German regimental markings. A number of these markings, together with a brief outline of how to interpret them, are found in the **Inscriptions in Western Languages** chapter. The reader familiar with the first edition will also notice the addition of chapters on German police markings and World War II German Mauser subcontractor codes.

Finally, I will take this opportunity to thank Mr. John Wall for his exceptional generosity in furnishing me, on his own initiative, considerable information on Czech Mauser markings. Along the same lines I must equally express my gratitude to Mr. Francis Allan for his kindness in sending me ---and a delightful surprise it was--- a copy of his and Mr. Roger Wakelam's invaluable treatise on Siamese Mausers. Thank you again, gentlemen.

Terence W. Lapin
Arlington, VA
November 2001

MAUSER MILITARY RIFLE MARKINGS

ORGANZIATION OF THE BOOK

I arranged the identification chapters and tables of this book into categories based on what I felt was a logical format. The chapters having inscriptions in various languages are divided into two parts: "**Inscriptions in Western Languages**" seemed to me to be sufficiently descriptive for most non-English words in Roman letters, and "**Inscriptions in Non-Western Languages**" is a convenient way to describe the Arabic, Chinese, Cyrillic, and other non-Roman writing systems. "**German World War II Codes**" and similar titles are self-explanatory.

Anyone trying to identify "**Coats of Arms**" or things resembling them would, I hope, head right for the chapter thus titled, although there is a sufficient difference, to my way of thinking, between "coats of arms" and many other illustrations. These other illustrations I have called "**Emblems**", and if the reader does not find what he is hunting for in one chapter the quarry may well be found in the other.

As chapter titles, "**Numbers**" "**Calendars and Dating**", and so forth are as clear as I thought possible. Having come to these conclusions I simply arranged the various items alphabetically, usually by country, with the thought that such was a reasonable layout.

I trust that this arrangement will prove convenient and easy for the reader.

MAUSER MILITARY RIFLE MARKINGS

LANGUAGES

Because Mauser rifles have been made for countries all over the world is it logical that they should contain markings in many different languages. Some of these, such as German, French and Spanish, are relatively familiar to English-speaking readers and thus are easy to deal with; these are, I feel, properly covered in the **Inscriptions in Western Languages** tables of this book; Czech and Turkish, so different from English but not really "foreign" in the sense of having a different writing system, require their own discussion here.

In the present chapter I discuss some of the characteristics of the more "exotic" Mauser languages so that the items in the **Inscriptions in Non-Western Languages** chapter will be more informative and understandable. There are only a few of these languages, and I hope the reader will bear with me here and even find something of interest.

ARABIC ---

Ironically, this language usually appears on Mauser rifles not in its own right, but mostly in the form of two of the numerous other languages that use Arabic-derived scripts (Persian and Ottoman Turkish, which are covered later in this section). Arabic is written with the letters linked, as handwriting is in English; the concept of individual printed letters has never caught on for Arabic.

I did not include a complete table of the Arabic alphabet in the **Alphabets** chapter of this book for a good reason: almost all of the letters change their form depending on whether they are the first, last, or middle letter in a word, or stand alone; also, when some letters are combined with another letter in a ligature ---which is very often--- they are difficult if not impossible to distinguish unless you have a fairly good knowledge of the language. A simple table of the alphabet is of no benefit in reading an Arabic inscription: you have to know *how* the letters are written and connected. That being said, as individual letters do appear occasionally, especially in serial numbers, I have included a table of the independent forms of the letters in the **Alphabets** chapter.

Short vowels are generally not written in Arabic except in religious texts and where on the rare occasion they are necessary to avoid an important ambiguity.

Such Arabic as appears on Mausers is usually quite tiny as it is mostly on national crests found on the receivers. An exception to this generalization is the letter *jeem*, the fifth letter of the Arabic alphabet, which is common on Iraqi Mausers. I believe this to be an abbreviation for the word *jaysh*, meaning "army"(see the **Emblems** chapter). Arabic numerals are covered in the **Numbers** chapter.

CHINESE ---

This language can be written ---both words and numbers--- left to right, right to left, or top to bottom; you have to determine the direction from the context of the characters. The more common lateral way of writing during the Mauser period was right to left, but there are several exceptions. Nowadays it is most often written left to right.

Classical Chinese does not have an alphabet, although there are several major systems to transcribe it into Latin letters. The government of the People's Republic of China developed an official system for using Latin letters to write Chinese, called *pinyin*, in the late 1950s, but this does not concern us as it does not appear on Mausers.

MAUSER MILITARY RIFLE MARKINGS

Chinese is written in characters called *ideographs*, which originally were pictures of things they represented but developed into characters for sounds or concepts; now they generally have no relation to what they originally depicted. As there are more than 50,000 ideographs, and about 5,000 are used in daily newspapers, there is a sufficiently good reason that I did not list them in this book. I will mention that since the 1950s the government of the PRC has pursued a campaign of simplifying Chinese ideographs by eliminating some strokes and otherwise making them easier to read and write. The Republic of China (Taiwan) and traditional Chinese communities overseas continue to use the old style characters. The simplified characters are in many instances so different from the old ones that sometimes people who know one form of an ideograph fail to recognize it in its other form.

The Chinese inscriptions on Mausers are depicted in the **Inscriptions in Non-Western Languages** charts. Chinese numerals are discussed in the **Numbers** chapter.

CZECH ---

Czech and Slovak, the main languages of ex-Czechoslovakia, are not the same language, though they are closely related; each has some letters the other does not have, and there are other differences. For our Mauser purposes we need consider only Czech.

The stress in Czech is always on the first syllable, and I mention this only because Czech vowels can take what appears to be an "accent" mark, becoming á,é,í,ó,ú, and ý. This slash-like mark (called *čárka* in Czech, by the way) indicates that the vowel it sits on is pronounced about twice as long as one without it, so *á* represents a long form of *a*. Thus, *statná*, "state" ---which appears in a Mauser inscription shown in the **Inscriptions in Western Languages** chapter--- is pronounced "STAT-naa", not "stat-NA".

Besides the long vowels there are other Czech letters with diacritical marks (those dots, circles, and other adornments that English has largely managed to avoid), namely č, d', ě, ň, ř, š, t', ů, and ž. Because it occurs so frequently in the Czech Mauser inscriptions I will note here that č is pronounced like the *ch* in "chick". That little "v" on top of the letter, by the way, is called a *háček*. For the sake of completeness I should mention that the ° on ů is called a *krouzek*; ú is written ů when it is not the first letter in a word.

Before leaving Czech I will note that, like many languages, it has a number of cases ---seven, actually; English, by way of comparison, has two--- meaning that the ending of a Czech word must change according to the word's use. For this reason the place name *Brno* appears as *Brně* in those Mauser inscriptions in which it follows *v* (a preposition meaning "in"); Brně has the "locative" (also called the "prepositional") case ending Brno needs to mean "in Brno".

ETHIOPIAN ---

Though a number of tongues are spoken in Ethiopia the national language is Amharic, and that is what appears on the Ethiopian Mauser.

Amharic (the language of the Amharas, the predominant ethnic group) is distantly related to Arabic, Hebrew, and other Semitic languages. The alphabet used for this language is not native to the country but derives from one that originated in southern Arabia about 2,000 years ago, where it was used for the local Arabic dialects; it came to Africa through trade and war. The written predecessor of Amharic, called *Ge'ez*, is still used as a liturgical language in the Ethiopian Orthodox Church. The Ge'ez alphabet and numerals are slightly different from Amharic, and appear in part of the inscription shown in the **Inscriptions in Non-Western Languages** section. There is no English equivalent for the relationship between Ge'ez and Amharic; for those familiar with the Slavic languages it is comparable to the relationship between Old Church Slavonic and Russian.

Amharic script is actually more a syllabary than an alphabet as each of the 37 letters has an intrinsic vowel sound associated with it, so "j" is not really "j" but "ja", "ji", "jo", etc. There are 247 possible combinations, each involving a slightly different writing of each letter. Once more, I judged that including an alphabet table would be impractical. The rifle inscriptions are illustrated and translated in the **Inscriptions in Non-Western Languages** chapter. Ethiopian numerals are covered in the **Numbers** chapter.

HEBREW ---

Like its cousin Arabic, Hebrew is written from right to left, though the two languages' alphabets are otherwise completely different. The Hebrew alphabet has 27 letters, five of which change their form depending on where they appear in a word. As with Arabic and Persian, short vowels are very seldom written except in religious texts. Generally, the only Hebrew found on Mauser rifles is in the Israel Defense Force crest and as a one- or two-letter proof mark. A copy of the Hebrew alphabet is included in the **Alphabet** chapter. Hebrew letters are also sometimes used in writing dates, and letters with numerical values are listed in a chart in the **Numbers** chapter. In conjunction with this the reader is advised to consult the **Calendars and Dating** chapter as well.

PERSIAN ---

"Persian" in Persian is "Farsi", originally the language of the region called Fars, in what is now southwestern Iran. I refrain from calling it *"Farsi"* for the same reason that I do not feel compelled to call Romanian *limba Româna*, Norwegian *Norsk,* or Turkish *Türkçe*.

Persian is an Indo-European language, as are English, Spanish, Russian, and Albanian, among many others. It is written from right to left in a modified form of the Arabic alphabet which includes four additional letters for sounds that do not exist in standard Arabic ("p", "ch", "zh", and "g"). Moreover, many Arabic letters are pronounced quite differently in Persian. Short vowels are even less commonly written in Persian than in Arabic. The usual style of script used in Persian is *nasta'liq*, and that is what appears on the Persian Mausers.

The beautiful imperial Persian crest is described in the **Emblems** chapter, the inscriptions are set forth in the **Inscriptions in Non-Western Languages** chapter, and the numerals are covered in the **Numbers** chapter.

SERBO-CROATIAN ---

Serbian and Croatian are essentially just dialects of the same language, spoken in what used to be Yugoslavia where they had official status along with two other Slavic tongues, Macedonian and Slovene.

MAUSER MILITARY RIFLE MARKINGS

Serbian is written in a form of the Cyrillic alphabet, which is itself derived from the Greek alphabet. Creation of the Cyrillic alphabet is traditionally attributed to the brothers St. Cyril (hence "Cyrillic") and St. Methodius --- though there is considerable academic dispute about this--- who brought Christianity to the Slavs beginning in c. 863 A.D. Most Serbian letters have the same sound values as in the similar Cyrillic alphabet of Russian; there are six letters which do not appear in Russian Cyrillic, and which are peculiar to Serbian Cyrillic.

Croatian is written in Latin letters with seven additional letters for sounds not found in the Latin alphabet.

Both alphabets appear in Yugoslav Mauser inscriptions. Croatian uses Latin letters and Serbian uses Cyrillic for the same reason Polish and Czech are written in Latin letters, and Russian and Ukrainian in Cyrillic: the Slavic peoples who adopted Roman Catholic Christianity use Latin letters and Eastern Orthodox Christians use Cyrillic.

SIAMESE (THAI) ---

The Thai people and their language originated in what is now southern China, though the language is not a Chinese dialect. Their alphabet, however, derives from those of southern India. Thai letters, like the numbers, are read from left to right.

The Thai alphabet has 44 letters consisting of a consonant with a built-in vowel sound of o, 22 vowel signs, 3 more combinations to express double vowels, and another sign indicating that a sound is nasalized. There are also 4 tone markers: Thai is a tonal language, and the meaning of a word depends on which of five tones it is spoken with; one of the tones does not have a marker. Thai writing does not use breaks between words: they run together until the writer decides to change the thought they express.

All this makes for a rather complicated writing system, but one typical of the languages of southern India.

The word *Thai* means "free" in the Thai language; the Siamese government changed the country's name to *Prathet Thai* (Land of the Free; "Thailand", to us) in 1938, but the older name, Siam, hung on, especially among the British and people who like Siamese cats or *The King and I*.

TURKISH ---

Turkish is a language of Central Asian origin, very distantly related to Hungarian, Estonian and Finnish. Before the language reforms of 1928 Turkish was written in a modified form of the Arabic alphabet quite similar to that used for Persian, and for the same reasons. The Arabic alphabet was unsuited to the sounds ---especially the vowels--- and grammatical structure of the Turkish language but was used because of the Turks' long and close religious and cultural association with Arabs and Persians.

In the late 1860s a commission was appointed by Sultan Abdülaziz to study a proposed adoption of the Latin alphabet to write Turkish. The commission rejected the idea. One of the commission's scholars declared that writing from left to right was "uncomfortable and stupid", and the idea was shelved for more than sixty years until Kemal Atatürk, founder of the Republic, simply ordered it to be implemented.

Modern Turkish is written in a modified Latin alphabet containing several letters for sounds peculiar to Turkish; in addition, some Latin letters are pronounced

MAUSER MILITARY RIFLE MARKINGS

differently in Turkish than they are in other languages. See the table in the **Alphabets** chapter.

Since the 1920s there has been a government undertaking to rid Turkish of the huge number of Arabic and Persian loan words in Ottoman Turkish and replace them with "real" Turkish words, no matter how archaic. A middle-aged Turk today would have considerable difficulty understanding the spoken language of his great-grandparents, and cannot read it at all unless he has studied it as, to all intents and purposes, a foreign language.

The term "Ottoman", by the way, refers to things Turkish of the imperial era. The word is a corruption of the Turkish *Osmanlı*, which comes from the name of Osman, the founder of the Ottoman dynasty.

The favored style of script for official inscriptions in Ottoman Turkish was *riq'a*, and that is what usually appears on the rifles, though the Persian-influenced *nasta'liq* script is used for the Customs Administration marking on the receiver ring. Although the Empire was abolished and a republic declared in 1922, use of the old alphabet continued until introduction of the modified Latin alphabet in 1928.

Firearms enthusiasts may be interested in one English word derived from Ottoman Turkish. The Arabic term *dār as-sinā'a*, "workshop" (literally, "abode of industry") had, by the late Middle Ages, acquired the narrower meaning of "military shipyard". One Ottoman shipyard (Algiers?) was also the site of a major arms works. The Turks borrowed the Arabic words and pronounced them in the Turkish way, melding them into one word as they usually did. This word was in turn picked up by Italian and Spanish merchants and imported into Europe, where it became *arsenal*. To this day the primary meaning of *arsenal* in Spanish (and a secondary meaning in Italian) is "dock" or "shipyard".

MAUSER MILITARY RIFLE MARKINGS

ALPHABETS

The following are some of the more common foreign alphabets used in Mauser markings. Generally, the English equivalents given are approximations, as some of the foreign sounds do not exist in English at all; these are marked N/E for "no equivalent". For a discussion of related matters see the chapters on **Languages** and **Numbers**.

ARABIC	ENGLISH	ARABIC	ENGLISH
ا	No sound*	ض	D in *dull*
ب	B	ط	T in *tot*
ت	T	ظ	TH in *thud*
ث	TH in *thin*	ع	N/E **
ج	J	غ	N/E ***
ح	A harsh *H*	ف	F
خ	CH in *Bach*	ق	N/E ****
د	D	ك	K
ذ	TH in *these*	ل	L
ر	R	م	M
ز	Z	ن	N
س	S in sin	ه	H
ش	SH	و	W or Ū
ص	S in *sod*	ى	Y, Ā or Ī

* Sometimes this is like the *a* in *father*, sometimes it is just the silent "chair" for a short vowel (*a*, *i* or *u*) to "sit" on.
** There is nothing in English approaching this sound other than that produced by a gag reflex.
*** This is a grating sound similar to a French *r*.
**** Usually transliterated as *q*, this is a *k* sound produced at the back of the throat.

MAUSER MILITARY RIFLE MARKINGS

The following table shows German *Fraktur* letters, which are very common on Mausers made before the end of the First World War. Note that the letters are similar, but not identical, to "Old English" lettering. The Germans call this type *Fraktur* because of the "fractured" appearance of the letters. Some pairs of *Fraktur* letters are very similar in appearance, for example B and V; I and J; O and Q.

Before the 1940s "serious" books, newspapers, and other publications in Germany were printed in *Fraktur* letters. During the early part of World War II the German government began to expand its use of Roman letters, especially in German-language publications intended for distribution outside Germany, as a means to encourage the study of German by the subjugated peoples of the occupied territories of northern and western Europe. The use of *Fraktur* has declined steadily since the end of the war.

German	Roman	German	Roman
𝔄	A	𝔑	N
𝔅	B	𝔒	O
ℭ	C	𝔓	P
𝔇	D	𝔔	Q
𝔈	E	𝔑	R
𝔉	F	𝔖	S
𝔊	G	𝔗	T
ℌ	H	𝔘	U
ℑ	I	𝔙	V
ℑ	J	𝔚	W
𝔎	K	𝔛	X
𝔏	L	𝔜	Y
𝔐	M	𝔷	Z

MAUSER MILITARY RIFLE MARKINGS

There is also a double letter ß, *ss* (derived from an old written form of *sz*); it is used only with some very specific spelling rules and even then only in lower-case form.

When two forms of a letter are given in the Hebrew Table the one on the left is a form used at the end of a word.

Hebrew	English	Hebrew	English
א	No sound	ם מ	M
ב	B	ן נ	N
ב	V	ס	S
ג	G	ע	No sound
ד	D	פ	P
ה	H	ף פ	F
ו	V	ץ צ	TS in *cats*
ז	Z	ק	K
ח	CH in *Bach*	ר	A French *R*
ט	T	שׁ	SH
י	Y	שׂ	S
ך כ	K	ת	T
ך כ	CH in *Bach*	ת	T
ל	L		

Persian contains four letters not found in the Arabic alphabet; these were also used in Ottoman Turkish and are shown in the table below.

PERSIAN	ENGLISH
پ	P
چ	CH in *church*
ژ	S in *measure*
گ	G in *go*

MAUSER MILITARY RIFLE MARKINGS

Serbian Cyrillic	English
А	A as in *father*
Б	B
В	V
Г	G as in *go*
Д	D
Ђ	Almost like the *j* in *jack*
Е	E as in *end*
Ж	Like the *s* in *measure*
З	Z
И	EE as in *seen*
Ј	Y as in *you*
К	K
Л	L
Љ	Rather like *lli* in *million*
М	M
Н	N
Њ	*Ny* as in *canyon*
О	O
П	P
Р	R
С	S
Т	T
Ћ	Almost like *ch* in *church*
У	OO in *moon*
Ф	F
Х	CH as in *Bach*
Ц	TS as in *cats*
Ч	CH as in *church*
Џ	Almost like *j* in *jeep*
Ш	SH as in *sharp*

MAUSER MILITARY RIFLE MARKINGS

Modern Turkish	English
A	A as in *father*
B	B
C	J as in *jeep*
Ç	CH as in *church*
D	D
E	E in *let* or *ay* in *lay*
F	F
G	G as in *go*
Ğ	No sound; it lengthens the preceding vowel
H	H
I	Similar to the *u* in *hug*
İ	I in *pit*
J	Like the *s* in *measure*
K	K
L	L
M	M
N	N
O	O
Ö	Like German *ö* or French *eu*
P	P
R	R
S	S
Ş	SH in *shine*
T	T
U	OO as in *moon*
Ü	Like German *ü* or French *u*
V	V
Y	Y as in *yam*
Z	Z

MAUSER MILITARY RIFLE MARKINGS

SLASHES, HOOKS, DOTS AND SO ON

In the **Inscriptions in Western Languages** chapter and elsewhere the reader will see a number of marks on, over, and under some of the letters in some foreign words. English has been largely spared these things (the technical term is "diacritical marks"), for which we should be truly thankful. My purpose in mentioning them is to note that, as to the languages used on Mausers, the only ones in which these marks actually indicate where to stress a word's pronunciation, i.e., "accent" a syllable, are Spanish and Portugese.* The *á* in the Czech word for "state", *statná*, for example, does **not** mean the word is pronounced "stat-NAH": all Czech words are stressed on the first syllable, and this *á* has an "accent" mark to indicate that it represents a sound different from that of the unmarked Czech *a*. For more about this see the discussion of Czech in the **Languages** chapter.

There are many different diacritical marks, representing a large number of variables; they are, however, outside the scope of this book and I mention the matter only to avoid confusion to readers unfamiliar with some of these languages and their writing systems.

---

* Even with these two languages there are exceptions to this rule; however, as none of these exceptions appear on Mausers they need not concern us.

MAUSER MILITARY RIFLE MARKINGS

NUMBERS

Foreign numbers found on military Mauser rifles include those from the Amharic (Ethiopian), Arabic, Chinese, Hebrew, Persian, and Thai languages.

AMHARIC (ETHIOPIAN) ---

The numbers, like the language, are written left to right. Numbers greater than 10 are combined as follows: 19 is written 10 9; 37 as 30 7, etc.; 1926 is 10 9 100 20 6, and so on.

δ̄	B̄	Γ̄	Ō	c̄	ȷ̄	Z̄	Ī	H̄	ī
1	2	3	4	5	6	7	8	9	10
X̄	Ū	Ȳ	Ȳ	Ȳ	C̄	T̄	Ȳ	Q̄	♦
20	30	40	50	60	70	80	90	100	0

ARABIC, PERSIAN and TURKISH---

Although Arabic writing and its derivatives used for Persian and Ottoman Turkish are read from right to left, the numbers appear to be written from left to right, as in English. This is because of the way in which numbers in the Arabic language are *spoken*: an Arabic speaker says, for example, the number 24 as "four and twenty"(*arba'a wa 'ishrīn*); consequently, the numeral 4 is written before the numeral 2 of "20": ٢٤. The fact that the numbers *seem* to be written left to right, as in English, is merely a coincidence. In earlier times the entire number was read

MAUSER MILITARY RIFLE MARKINGS

right to left, but modern practice is to read or say the largest unit first and work backward, using the "four-and-twenty" form only for the numbers from 21 through 99; however, the largest unit is always *written* last, i.e., the first on the left. For example, the date 1329 is written ١٣٢٩ but read as "one thousand three hundred nine and twenty".

In Persian and Turkish, however, the numbers are spoken in the same order as in English: to use the same example, 24 in Persian is said "twenty and four" (*bist o chahar*), and in Turkish it is "twenty four" (*yirmi dört*). Even though the numbers are written "backwards" for a Turkish- or Persian-speaker, Persian continues to write the numerals in the same order as Arabic simply as a matter of longstanding tradition, as did Turkish before the language reforms of 1928.

One often sees Arabic numbers referred to as "Farsi". This designation is wrong: the two languages use numerals sufficiently different to warrant a distinction between them. Persian or "Farsi" (see the **Languages** chapter elsewhere in this book for a discussion of this matter) uses several archaic forms of numerals that the Arabs discarded many centuries ago: specifically the way of writing 4, 5 and 6. From around 1000 A.D. until about 1600 the Turks used the Persian forms of the numerals; thereafter they used the more "modern" Arabic numbers until 1928.

The Arabs acquired their numerals from India, changing the forms of almost all of them and reassigning some of the values. For example, the numerals 2, 3, and 8 got tilted 90° clockwise, 9 got reversed, and the Indian 0 became the Arab 5. The West adopted these numbers from the Arabs during the late Middle Ages, replacing the much more awkward Roman numerals which had been the Western system of writing numbers for over a thousand years.

MAUSER MILITARY RIFLE MARKINGS

Strangely, the numerals we use now look much more like the original Indian figures than those first transmitted to Europe through the Arabs. The author learned them as "Hindu-Arabic" numerals in school; that term is probably politically incorrect now and, if so, would prove that it is the right one.

ARABIC AND OTTOMAN TURKISH NUMERALS

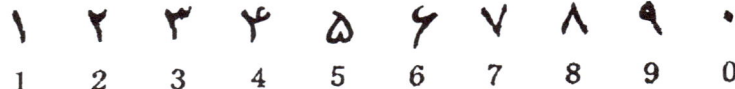

PERSIAN NUMERALS

۱	۲	۳	۴	۵	۶	۷	۸	۹	۰
1	2	3	4	5	6	7	8	9	0

CHINESE---

Numbers in Chinese ---like the words--- can be written right to left, left to right, or top to bottom. This occasionally causes problems in reading them, but this can be overcome by looking for clues to the numbers' orientation.

Traditional Chinese numbers did not include a sign for zero, although occasionally one of at least two different native Chinese ideographs was used; the Western numeral was borrowed in the 20th century. To compensate for the lack of zero Chinese numeration generally indicated the comparable number by use of the character for 10. Thus,

MAUSER MILITARY RIFLE MARKINGS

20 is written 二 十, i.e., 2 [x] 10, and 25 is 二十五, i.e., 2 [x] 10 [+] 5. The problem here is that the numerals can be read in reverse, in which case the number in the first example would be read as 12 (i.e., 10 + 2) and the second as 52 (i.e., [5 x 10] + 2). Context is the key to getting them right.

When numerals are found on Chinese Mausers they are generally in the traditional form. The tendency over the past several decades, especially in the People's Republic of China, has been to use Western numbers, and this trend is likely to increase in all parts of the Chinese-speaking world.

一	二	三	四	五	六	七	八	九	十	百
1	2	3	4	5	6	7	8	9	10	100

HEBREW ---

א	ב	ג	ד	ה	ו	ז	ח	ט	י	כ
1	2	3	4	5	6	7	8	9	10	20
ל	מ	נ	ס	ע	פ	צ	ק	ר	ש	ת
30	40	50	60	70	80	90	100	200	300	400

Like the script, Hebrew numerals are written right to left. Note that in writing dates Hebrew does not use thousands. Note also that the largest number is 400, and that numbers greater than that are written as combinations of the numerals for lesser numbers; for example, 685 is written ת ר פ ה, i.e., 400 and 200 and 80 and 5. Occasionally the letter/numeral ״ (0) will be placed before (i.e., to the right of) a digit for a number from 1 through 9 in a date, though this is somewhat rare.

MAUSER MILITARY RIFLE MARKINGS

SIAMESE (THAI) ---

      Like Thai writing the numerals are written left to right. Thai numbers are combined as they are in Western numerals, so 12 is written ๑ ๒, 70 is ๗ ๐, and so forth.

      Note the use of zero: like the letters, Thai numerals are derived from writing systems of southern India; so, too, the concept of zero as a numerical placeholder is a borrowing from India. The Maya of Central America developed the concept of zero independently, but they exerted no influence in its use as their civilization collapsed even before the arrival of the Spanish and the idea was already established in Europe from trade with the Arabs.

๑	๒	๓	๔	๕	๖	๗	๘	๙	๐
1	2	3	4	5	6	7	8	9	0

MAUSER MILITARY RIFLE MARKINGS

CALENDARS AND DATING

Because of the worldwide distribution of Mauser rifles it is not surprising that one finds different dating systems on various models. It is useful to take note of these calendar systems; fortunately, there are fewer than a half-dozen with which we need concern ourselves.

The Chinese Calendar ---

Before the founding of the Republic of China in 1911 (on the "Double 10", October 10th) the Chinese reckoned their calendar by a complex combination of the reigns of emperors and regularly recurring cycles. This can be a nightmare for conversion to A.D. dates but, fortunately, the matter does burden us much as only two emperors reigned during the Mauser era: Tsai-tien (throne name: Kuang-hsü), whose reign was 1875-1908; and his nephew and successor Hsüan-tung (more commonly known as Henry Pu-yi), who reigned from 1908 until his forced abdication in 1912. The hapless Pu-yi made an unfortunate comeback as emperor of the Japanese puppet state of Manchukuo in the 1930s and '40s. After spending the post-war years in a Chinese prison for war criminals Pu-yi was released in the late 1950s and ended his days, in 1967, as a minor bureaucrat ---not a gardener, as fiction has it--- in a government horticultural office in Beijing.

We do need to note the post-Revolution Chinese practice of dating from the year 1911. This is reasonably simple as one just adds the one- or two-digit post-Revolution year to 1911. For example: a Chinese rifle dated for the year "21" is equivalent to 1932 A.D. This practice was discontinued on the Mainland after the Communist victory in 1949, but continues to this day for some purposes in the Republic of China (i.e., Taiwan.) See also the **Notes**

MAUSER MILITARY RIFLE MARKINGS

on China in the **Inscriptions in Non-Western Languages** chapter.

The Ethiopian Calendar---

Ethiopia takes great pride in being the first Christian country (though this distinction is disputed by the Armenians and the Georgians). In any event, Ethiopia uses a unique dating system that is 7 years and 8 months behind the Gregorian A.D. year used by most of the rest of the world. To obtain the A.D. year add 8 to the Ethiopian year. For example, the Ethiopian year 1926 is 1934 A.D.

The Ethiopian system to write the year as a date uses numerals in the following way: the year 1926, for example, is written 10 9 100 20 6, i.e., (10+9) x 100 + (20+6). Sometimes the date is followed by a one-letter abbreviation for "year".

The Islamic Calendar---

This system reckons dates from 622 A.D. when the Prophet Muhammad emigrated from Mecca to Madina, in what is now Saudi Arabia. It is called the *hijri* calendar, from the Arabic word *hijra*, meaning emigration; its dates are usually noted as "A.H." in Western publications, the abbreviation for *anno Hegirae* (Latin for "year of the Hegira, or Hijra") to make it analogous to the use of A.D.

This dating system was common to almost the entire Islamic world for centuries. Divergences appeared over the years, including a major break by Persia in 1925 (which is discussed below), and by Turkey at the end of that same year. The Islamic calendar is based on a lunar year, rather than a solar year like the Gregorian calendar (now commonly used all over the world) and the old Julian calendar. Lunar months and solar months differ in the number of days per month, making for a shorter calendar in the lunar year; moreover, New Year's Day moves around the calendar in the Islamic year. Consequently, the two years, A.D. and A.H., do not coincide. The formula to convert an A.H. year to an A.D. year is as follows: multiply the A.H. date by .03, round that product, and add 622.

Example A: 1277 A.H. x .03 = 38.31; 1277 - 38 = 1239; 1239 + 622 = 1861 A.D.
Example B: 1336 A.H. x .03 = 40.08; 1336 - 40 = 1296; 1296 + 622 = 1918 A.D.

This formula will give you either the exactly equivalent A.D. year or will be one year off. The overlap of years is unavoidable unless you resort to more sophisticated mathematics. For our purposes the foregoing formula is sufficient--- with practice you can even do it in your head. It is most unlikely (though not inconceivable) that one would know the exact day of the year a particular Mauser rifle was made and thus would want to know the precise date conversion; in that case you can go to any of the many calendar conversion sites on the Internet.

The most common series of Mauser rifles involving the Islamic calendar is that of Turkey, which used the Islamic calendar until adopting the Western (Gregorian) calendar on December 26, 1925. The Turks made a minor change to their calendar in 1917, which entailed a switch from "Greek" to Gregorian months ---but *not* to the Gregorian year--- but this does not affect the dating formula.

MAUSER MILITARY RIFLE MARKINGS

The Jewish Calendar ---

I have included this entry because of the possibility that one will encounter a Hebrew marking on an Israeli Mauser which appears to make no sense linguistically, and which may actually be a date. Hebrew letters have numerical values, and the Israelis commonly use them in writing dates. Hebrew letters and their numerical values are found in the **Numbers** chapter of this book. The Jewish calendar in use today is quite old, dating from about 360 A.D., and is based on the traditional date reckoned for the creation of the world. As with most other dating systems there is no exact correspondence with the A.D. system; the Jewish year moves around a bit, and New Year's Day, *Rosh Hashana* (literally, "head of the year"), occurs on varying days in September.

To get an approximate conversion from the Jewish year to A.D. subtract 3,760 (sometimes 3,761, because of the inexact correspondence of the two dating systems) from the Jewish date. As a benchmark for the reader to work with, the Jewish year 5708 = 1948 A.D., which is the year of the founding of the State of Israel.

Bear in mind that thousands are not expressed in these Jewish dates, and that the numbers are derived quite differently from the Western system. 5708 is written like this: ח״שת, which is "400 [and] 300 [and] 0 [and] 8", reading right to left, as in Hebrew script. Note that 0 is often omitted.

The Persian Calendar ---

Although it is a *very* Islamic country Iran (which officially changed its name from Persia in 1935) uses a different calendar from that of the rest of the Islamic world. The Iranian calendar year begins on the vernal equinox, on or about March 21st --- the first day of spring in the Northern Hemisphere; moreover, the Iranian year is solar, unlike the lunar calendar of other Muslim nations. Because the Western and Iranian years always begin on different days they overlap, hence there will be a possible difference of one year when an Iranian year date is converted to a Western one, and vice-versa. The formula given above for converting A.H. dates to A.D. dates does **not** apply to Iranian dates, especially during the "Mauser era", as Persia changed to the solar year in 1925, which suddenly went from 1344 to 1304.

To convert the Iranian year to A.D. from 1925 on, just add 622 to the Iranian date. Note that adding 622 to 1304 yields 1926 rather than 1925: this is because of the calendar overlap, as previously described; sometimes one must add 621 --- it depends on the particular day of the year.

Some readers may be interested to learn that the last shah of Iran, the late Mohammad Reza Pahlavi, tried to institute a bizarre dating system in 1976, which became the year 2535. This was derived by adding 2,500 years to his own reign to tie it into the founding of the ancient Persian Empire; the Shah's empire, however, went all the way back to his father, Reza, a cavalry officer who deposed the last shah of the Qājār Dynasty in 1925. The eccentric dating system lasted about as long as Mohammad Reza Shah, and was abandoned in 1978.

It is possible, though highly unlikely, that some ceremonial Mauser rifle may be found with such a 2535-2537 date, and I mention the last calendar change only for the sake of completeness.

MAUSER MILITARY RIFLE MARKINGS

The Siamese (Thai) Calendars ---

The Thai have several calendars, but to determine which one is being applied to the date you are reading requires only a slight bit of mental exercise. One calendar is based on the year of the Buddha's birth, reckoned as 543 B.C; this yields a year which, at present, will be in the 2500s. To convert the Thai Buddhist year to A.D. subtract 543 from the Thai year. For example, the 1923 Thai Mauser has the designation "Type 66" because it was adopted in the Thai Buddhist year 2466, which is 1923 A.D.: 2466 – 543 = 1923.

Another frequently used calendar is the Bangkok Era or *Ratanakosind-sok* calendar, dating from 1781. This is easily recognizable because the Thai date will have only three numerals. To convert a date from this calendar to A.D. add 1781 to the Bangkok Era date.

The third calendar is the *Chula-Sakarat*, which is of Burmese origin and began in 638 A.D. This will give a Thai date at present in, obviously, the 1300s. To convert a Chula-Sakarat date to A.D. add 638 to it.

MAUSER MILITARY RIFLE MARKINGS

INSCRIPTIONS IN WESTERN LANGUAGES

This chapter consists of inscriptions in Roman letters. For the reader's convenience they are arranged alphabetically by text rather than by country. Initials are listed before complete words. Inscriptions that begin with or are only numerals are listed last. Some individual letters can and do appear on rifles from various countries, but the reader should be able to attribute them to the correct nation easily by considering other markings and the rifle model itself. Bear in mind also that the same letter can be and often is used by the same country for a number of different meanings; for this reason the reader is advised to check several of the chapters when there is doubt or confusion about the exact meaning of a single-letter inscription.

For ease in reading most inscriptions that appear on the rifles solely in upper case letters have been changed to a normal upper case and lower case combination.

The writings that have an asterisk by the country name are discussed under the country's heading in the **Notes** at the end of this chapter.

Several of the following abbreviations may or may not appear on Mauser rifles but do appear on their accoutrements. For the most parts, the items described as World War II German factory codes are those of subcontractors, found on rifle parts; the rifle manufacturer codes are listed separately in the chapters titled **German World War II Codes, German Police District Codes,** and **German World War II Subcontractor Codes**.

MAUSER MILITARY RIFLE MARKINGS

Text	Meaning	Country
A	*Automobilní prapory* (Mechanized battalion)	Czechoslovakia
A.	*Artillerie Regiment* (Artillery Regiment)	Germany
A.	*Abteil* (Section)	Germany
A.F.	*Fussartillerie-Regiment* (Foot Artillery Regiment)	Germany
A.G. (also AG and A.-G.)	*Aktiengesellschaft* (Joint-stock company)	Germany
AR	*Albertus Rex* (King Albert, king of Saxony 1873-1902)	Saxony
AS.FA	*Ankara Silah Fabrikası Atölyeleri* (Ankara Arsenal Workshops)	Turkey*
AZ	*Automobilní zbrojovka* (Automobile arsenal)	Czechoslovakia
𝔄𝔪𝔟𝔢𝔯𝔤	Amberg; a city, site of a German arsenal	Germany
Ankara	The capital of Turkey	Turkey

MAUSER MILITARY RIFLE MARKINGS

Text	Meaning	Country
Armée d'Haiti	Haitian Army	Haiti
Armeria F.A. Rep. Dom.	*Armeria de las Fuerzas Armadas de la Republica Dominicana* (Arsenal of the Armed Forces of the Dominican Republic)	Dominican Republic
Avstånd	Range (distance)	Sweden*
B.A.	*Bekleidungsamt* (Clothing supply unit)	Germany
Bö	*Böhler* (made from Böhler Brothers factory steel)	Germany
BS	*Berlin-Spandau* (see Notes)	Germany*
BSW	*Berlin-Suhler Waffen- und Fahrzeugwerke* logo, only in 1937	Germany
B.T.	*Brücken-Train* (Bridging support group)	Germany*
BT3	Serbian Cyrillic for *VTZ*, abbr. for *Voini Tekhnichki Zavod* (Military Technical Factory)	Yugoslavia*

MAUSER MILITARY RIFLE MARKINGS

Text	Meaning	Country
Battalion Universitario de la Capital Federal	University Battalion of the Federal Capital	Argentina
Belgique	Belgium	Belgium
Brevet de W. & P. Mauser	W[ilhelm]. & P[aul]. Mauser's patent	Belgium
C	*Cyklistické prapory* (Bicycle battalion)	Czechoslovakia
ČETN	*Četnictvo* (Gendarmerie)	Czechoslovakia
C.F.S.	*Commando Federal de Seguridad* (Federal Security Command)	Argentina
C.G.H Suhl	C. G. Haenel (a manufacturer); Suhl (a German city)	Germany
CH.	*Chevauleger-Regiment* (Light-cavalry Regiment)	Bavaria
ČSK	*Československá* (Czechoslovakia)	Czechoslovakia
ČSZ	*Československá Statní Závody* (or *Zbrojovka*) (Czechoslovak State Factory [or Arsenal])	Czechoslovakia

MAUSER MILITARY RIFLE MARKINGS

Text	Meaning	Country
ČS. Závody na Výrobu Zbraní Brno	(ČS. = Československá) Czechoslovak Factory for Arms Production, Brno [a Czech city]	Czechoslovakia*
ČS. Zbrojovka Akc. Spol. v Brně	Československá Zbrojovka Akciový Společnost v Brně (Czechoslovak Arms Factory Joint-stock Co. in Brno)	Czechoslovakia
ČS.ST. Zbrojovka	Československá Statná Zbrojovka (Czechoslovak State Arms Factory)	Czechoslovakia
Československá Zbrojovka Brno	Czechoslovak Arms Factory, Brno	Czechoslovakia
Carl Gustafs Stads Gevärsfaktori	Carl Gustaf State Rifle Factory	Sweden
Chile Orden y Patria	Chile Harmony and Fatherland	Chile
D	Dělostřelecké pluky (Artillery Regiment)	Czechoslovakia

MAUSER MILITARY RIFLE MARKINGS

Text	Meaning	Country
D.	*Dragoner-Regiment* (Dragoon Regiment)	Germany*
D.G.F.M. – (F.M.A.P.)	*Dirección General de Fabricaciones Militares - (Fábrica Militar de Armas Portatiles)* (General Directorate of Military Production [Military Small Arms Factory])	Argentina
DOV	*Okresní doplňovací velitelství* (District Supplementary Command)	Czechoslovakia
DPLP	*Dělostřelecký protiletadlový pluk* (Anti-aircraft artillery battalion)	Czechoslovakia
DPS	*Divizní proviantní sklad* (Division Quartermaster Depot)	Czechoslovakia

MAUSER MILITARY RIFLE MARKINGS

Text	Meaning	Country
D.R.G.M.	*Deutsche Reichs Gebrauchsmuster* (German Empire Patent)	Germany
D.R.P.	*Deutsche Reichspost* (German Postal Service)	Germany
DSH	Code for *Výcvikový tábor Humenné* (Humenné Training Camp)	Czechoslovakia
DSj	Code for *Výcvikový tábor Jince-Čenkov* (Jince-Čenkov Training Camp)	Czechoslovakia
DSP	Code for *Výcvikový tábor Plavecké Podhradie* (Plavecké Podhradie Training Camp)	Czechoslovakia
DSV	Code for *Výcvikový tábor Vyškov* (Vyškov Training Camp)	Czechoslovakia
DZ	*Divizní zbojnice* (Division Arsenal)	Czechoslovakia

MAUSER MILITARY RIFLE MARKINGS

Text	Meaning	Country
Danzig	An ex-German city (now Gdańsk, Poland), site of an arsenal	Germany
Deutsche Waffen- und Munitions Fabriken	German Arms and Ammunition Factories	Germany
Deutsches Reich	German Empire	Germany*
E.	*Eisenbahn-Regiment* (Railroad Regiment); also for *Ersatz* (substitute)	Germany
E.A.	*Eisenbahn-Arbeiterkompanie* (Railway workers company)	Germany
E.B.	*Eisenbahn-Betriebskompanie* (Railway operation company)	Germany
EN	Ejército Nacional (National Army); an acceptance mark	Argentina, Nicaragua, Venezuela
E.U. do Brasil	*Estados Unidos do Brasil* (United States of Brazil)	Brazil

MAUSER MILITARY RIFLE MARKINGS

Text	Meaning	Country
E W B	*Einwohnerwehr Bayern* (Bavarian Paramilitary Police)	Germany (Bavaria)
Ejército Argentino	Argentine Army	Argentina
Ejército de Bolivia	Bolivian Army	Bolivia
Ejército de Colombia	Colombian Army	Colombia
Ejército del Ecuador	Army of Ecuador	Ecuador
Ejército Ecuadoriano	Ecuadorian Army	Ecuador
Eigentum Luftwaffe	Luftwaffe Property	Germany
Entubados a 7 m/m	[Re]barreled to 7 mm	Spain
Erfurt	A German city	Germany
Escuela Militar	Military Academy	Argentina
Escuela Naval	Naval Academy	Argentina
Espingarda Portugesa	Portugese Rifle	Portugal
Estados Unidos do Brasil	United States of Brazil	Brazil
État Belge	Belgian State	Belgium
F.	*Festung* (Fortress)	Germany
F.	*Fernsprecher-Abteilung* (Field telephone unit)	Germany

MAUSER MILITARY RIFLE MARKINGS

Text	Meaning	Country
F	Friedrich-Wilhelm IV (king of Prussia for 99 days in 1888)	Prussia
FA	Friedrich-August III (king of Saxony, 1904-18)	Saxony
FAO	*Fábrica de Armas de ymnas* (Oviedo Arms Factoryl)	Spain
Fab. Nat. D'Armes de Guerre	National War Arms Factory	Belgium
F.B. Radom	*Fabryka Broni* (Weapons Factory, Radom)	Poland
F.I.	*Fábrica de Itajubá* (Itajubá Factory)	Brazil
Fl.N.	*Flugüberwachung Bayern-Nord* ([Police] Air Surveillance Unit, North Bavaria)	Germany (Bavaria)
Fl.S.	*Flugüberwachung Bayern-Süd* ([Police] Air Surveillance Unit, South Bavaria)	Germany (Bavaria)

MAUSER MILITARY RIFLE MARKINGS

Text	Meaning	Country
F.N.	*Fabrique National* (National Factory)	Belgium*
FNRJ	*Federativna Narodna Republika Jugoslavija* (Federative Peoples' Republic of Yugoslavia)	Yugoslavia
FR. LANGENHAN ZELLA-MEHLIS	The Fr. Langenhan Co. of Zella-Mehlis (World War II subcontractor)	Germany
FS	*Finačni stráž* (Financial Guard)	Czechoslovakia
F.W.	*Friedrich Wilhelm* Monogram of King (later Kaiser) Wilhelm I	Germany*
Fábrica Checoslovaca de Armas S.A. Brno	Czechoslovak Arms Factory, Ltd., Brno (On arms for Spanish-speaking countries)	Czechoslovakia
Fábrica Nacional de Armas D.F.	National Arms Factory, *Districto Federal* (Federal District, i.e., Mexico City)	Mexico

MAUSER MILITARY RIFLE MARKINGS

Text	Meaning	Country
Fábrica de Armas de Oviedo	Oviedo Arms Factory	Spain
Fábrica de Itajubá – Brasil	Itajubá Factory – Brazil (Itajubá is a place)	Brazil
Fabrique d'Armes	Arms Factory	On Belgian rifles from FN
Fabrique Nationale	National Factory	Belgium
Fuerzas Militares	Armed Forces	Colombia
G	*Státní vojenské reformní reálné Gymnasium* (State Military Improved Secondary School), or *Vojenské gymnasium Moravská Třebová* (Military Secondary School in Moravská Třebová)	Czechoslovakia
G	On Kar98k receiver ring indicates rifle made in 1935	Germany
G. 24(t)	*Gewehr* (rifle) **t** = *tschechisch* = Czech. Vz. 1924, used by Germany	Germany*

38

MAUSER MILITARY RIFLE MARKINGS

Text	Meaning	Country
G. 220(b)	*Gewehr* (rifle) **b** = *belgisch* = Belgian. FN Model 1924, used by Germany	Germany*
G.29/40	*Gewehr* (rifle) model 1929, adopted by Germany in 1940	Germany*
G.33/40	*Gewehr* (rifle) Czech model 1933, adopted by Germany in 1940	Germany*
G. 41(M)	*Gewehr* (rifle) Model 1941 semi-automatic rifle	Germany
G.A.G.	*Grenz-Aufsichts-Gewehr* (Border Guards Rifle)	Germany
G.B.N.	*Gebrüder Bing, Nürnburg* (Bing Brothers, Nuremburg)	Germany*
Gew. 88	Gew. 88 *Gewehr* (Rifle) Model 1888	Germany
Gew. 91	Gew. 91 *Gewehr* Rifle Model 1891	Germany

39

MAUSER MILITARY RIFLE MARKINGS

Text	Meaning	Country
𝔊𝔢𝔴. 98	Gew. 98 *Gewehr* Rifle Model 1898	Germany
G.F.	*Garde-Fusilier* (Fusilier Guards, 1909 and later)	Germany*
G.F.R.	*Garde-Fusilier-Regiment* (Fusilier Guards, 1877-1909)	Germany
G.G.	*Garde-Grenadier* (Grenadier Guards, 1909 and later)	Germany*
G.G.R.	*Garde-Grenadier* (Grenadier Guards, 1877-1909)	Germany
G.J.	*Garde-Jäger-Bataillon* (Guards infantry battalion)	Germany
G.K.	*Garde-Kürassier* (Guards cavalry regiment)	Germany*
G.M.G.	*Garde-Maschinengewehr* (Guards machinegun company)	Germany
G.P.	*Garde-Pionier* (Guards engineer battalion)	Germany

MAUSER MILITARY RIFLE MARKINGS

Text	Meaning	Country
Geværfabriken, Otterup	Arms Factory, Otterup (a place in Denmark)	Denmark
Generalitat de Catalunya Comissió de L'Industria de Guerra Subsecretaria de Armament	Autonomous Government of Catalonia, War Industry Commission, Subsecretariat of Armaments	Spain
GmbH	*Gesellschaft mit beschränkter Haftung* (Limited liability company)	Any German-speaking country
Ginklų Fondas	Weapons Fund	Lithuania
Gobierno de Nicaragua	Government of Nicaragua	Nicaragua
H	Believed to indicate a rework at the *Reichswehr* facility in Hanover	Germany
H.	*Husaren-Regiment* (Light-cavalry Regiment)	Germany
HB	*Velitelství horských brigád* (Mountain Brigade Command)	Czechoslovakia

MAUSER MILITARY RIFLE MARKINGS

Text	Meaning	Country
HDB	*Velitelství hrubé dělostřelecké brigády* (Heavy artillery brigade HQ)	Czechoslovakia
HLS	*Hlavní letecký sklad* (Main Aviation Depot)	Czechoslovakia
H.L.Z.	*Hilfs-Lazarettzug* (Auxiliary field hospital unit)	Bavaria
HN	*Hraničářský prapor* (Border [Guard] Battalion)	Czechoslovakia
HP	*Hraničářský prapor* (Border [Guard] Battalion; added in 1939)	Czechoslovakia
HTS	*Hlavní telegrafní sklad* (Main Telegraph Depot)	Czechoslovakia
HVS	*Hlavní vozatajský sklad* (Main Towing Horse Units Depot)	Czechoslovakia
HZLS	*Hlavní železniční sklad* (Main Railroad Depot)	Czechoslovakia

MAUSER MILITARY RIFLE MARKINGS

Text	Meaning	Country
HZNS	*Hlavní železniční sklad* (Main Railroad Depot)	Czechoslovakia
HZS	*Hlavní zbrojní sklad* (Main Armory Depot)	Czechoslovakia
HÆR	Army	Norway
Husqvarna Vapenfabriks Aktiebolag	Husqvarna Arms Factory, Inc.	Sweden
J. G. Mod. 71	I.G. Mod. 71 *Infanterie Gewehr* (Infantry Rifle, Model [18]71)	Germany
I. S. S.	*Erste See Bataillon* (1st Sea Batallion)	Germany
K	On Kar98k receiver ring indicates manufacture in the year 1934	Germany
K	*Kraftfahrabteilung* ([Police] Motor Transport Unit)	Germany (Bavaria)
KAL. 22 LANG FÜR BÜCHSEN	Caliber .22 long rifle	Germany

MAUSER MILITARY RIFLE MARKINGS

Text	Meaning	Country
Karab. 98 b	Karab. 98b Abbr. For *Karabiner* (carbine) Model 98b	Germany
K.B.	*Kriminalpolizei Berlin* (Criminal Police, Berlin)	Germany
Kb.	Abbr. For *Karabin* (rifle)	Poland*
Kbk.	*Karabinek* (Carbine, i.e., a short rifle)	Poland*
K.I.	*Kadetteninstitut* (Cadet Academy)	Germany
K.K. Wehrsportgewehr	*Kleinkaliber-Wehrsportgewehr* (Small caliber Wehrmacht training rifle)	Germany
K.Kale	Abbr. for *Kırıkkale* (A suburb of Ankara, site of *Kırıkkale Tüfek Fabrikası*, a rifle factory)	Turkey
K.S.	*Kolonial Service* (Colonial Service)	Germany
K.S.	*Sonderwagenstaffel der Kraftfahrabteilung* ([Police] Armored Vehicle Squad)	Germany (Bavaria)

44

MAUSER MILITARY RIFLE MARKINGS

Text	Meaning	Country
L.	*Landwehr* (Provincial army reserves)	Germany*
L.	*Landjägerei* (Rural constabulary)	Germany
L.	*Landsturm* (Provincial third-line militia)	Germany*
Ldst.	*Landsturm* (Provincial third-line militia)	Germany
L.A.	*Luftschiffer-Abteilung* (Airships detachment)	Germany
L.B.	*Luftüberwachung Berlin* (Air surveillance, Berlin)	Germany
L.H.	*Luftüberwachung Hannover* (Air surveillance, Hannover)	Germany
L.He.	*Luftüberwachung Hessen-Nassau* (Air surveillance, Hessen-Nassau)	Germany
L.N.	*Luftüberwachung Niederschlesien* (Air surveillance, Lower Saxony)	Germany

45

MAUSER MILITARY RIFLE MARKINGS

Text	Meaning	Country
L.O.	*Luftüberwachung Ostpreussen* (Air surveillance, East Prussia)	Germany
L.P.	*Luftüberwachung Pommern* (Air surveillance, Pomerania)	Germany
L.S.	*Luftüberwachung Sachsen* (Air surveillance, Saxony)	Germany
LS	*Letecký sklad* (Aviation Depot; added in 1939)	Czechoslovakia
LS.Al.	*Landjägereischule Allenstein* (Rural Constabulary School, Allenstein)	Germany
LST.	*Landjägereischule Trier* (Rural Constabulary School, Trier)	Germany
L.W.	*Luftüberwachung Westfalen* (Air surveillance, Westphalia)	Germany

MAUSER MILITARY RIFLE MARKINGS

Text	Meaning	Country
La Coruña	A city in northern Spain, site of an arsenal	Spain
La Juventud es la Esperanza de la Patria	The Youth are the Hope of the Fatherland	Argentina
M	*Dělostřelecké měřičská náhradní rota* (Artillery measuring reserve company)	Czechoslovakia
M.	*Munitionskolonne* (Munitions column)	Germany
M.G.A.	*Maschinengewehr-Abteilung* (Machinegun unit)	Germany
M.G.K.	*Maschinengewehr-Kompagnie* (Machinegun Company)	Germany
M.S.	*Militär-Schiess-Schule* (Military Marksmanship School)	Germany
MT	*Muniční továrna* (Munitions factory)	Czechoslovakia
Manufacture d'Armes de l'État	State Arms Factory	Belgium

MAUSER MILITARY RIFLE MARKINGS

Text	Meaning	Country
Mo.	*Modelo* (Model)	Any Spanish-speaking country
Mod. Mauser	Model Mauser	South Africa (Boer republics)
MOD.24L.	Model 24 L.	Lithuania
MOD. 1922	Model 1922 carbine	Belgian-made for Brazil
Manufactura Loewe Berlin	Loewe Factory, Berlin	Spain
Mauser Chileno Manufactura Loewe Berlin	Chilean Mauser Loewe Factory, Berlin	Chile
Mauser Español Modelo 1893	Spanish Mauser Model 1893	Spain
Modelo	Model	Any Spanish- or Portugese-speaking country
Modelo Argentino	Argentine model	Argentina
N	*Nachrichten-technische Abteilung* ([Police] Technical Communications Unit)	Germany (Bavaria)
N.A.	*Ausbildungsstaffel der Nachrichten-technische Abteilung* ([Police] Technical Communications Training Squad)	Germany (Bavaria)

MAUSER MILITARY RIFLE MARKINGS

Text	Meaning	Country
N.Ba.	*Baustaffel der Nachrichten-technische Abteilung* ([Police] Technical Communications Maintenance Squad)	Germany (Bavaria)
NG	*Národní Garda* (National Guard)	Czechoslovakia
N.S.D.A.P.	*Nationalsozialistische Deutsche Arbeiterpartei* (National Socialist German Worker's Party; the Nazi party)	Germany
NUR FÜR KURZ PATRONE	Only for Short Cartridge (i.e., the 7.9x33 mm round)	Germany
OS	*Hlavní oděvní a lůžkový sklad* (Main Uniform and Bedding Depot)	Czechoslovakia
O.V.S.	*Oranje Vrij Staat* (Orange Free State)	South Africa*
Överslag	Impact above	Sweden*
P	*Pěší pluk* (Infantry Regiment)	Czechoslovakia

MAUSER MILITARY RIFLE MARKINGS

Text	Meaning	Country
P.	*Pionierbataillon* (Engineer Batallion)	Germany
P [serial number]	On rifles dated 1937 on the receiver ring, these are Czech-made vz 24s sold to China	China
PB	*Velitelství pěších brigád* (Infantry brigades HQ)	Czechoslovakia
PB.	*Polizeischule Bonn* (Police School, Bonn)	Germany
PD	*Velitelství polních dělostřeleckých brigád* (Field artillery brigades HQ)	Czechoslovakia
PDB	*Velitelství pěších divizí* (Infantry division HQ)	Czechoslovakia
P.F.K. Warszawa	*Pańswtowa Fabryka Karabinów* (Government Rifle Factory, Warsaw)	Poland

MAUSER MILITARY RIFLE MARKINGS

Text	Meaning	Country
P.H.	*Polizeischule Hannover* (Police School, Hannover)	Germany
P.He.	*Polizeischule Hessen-Nassau* (Police School, Hessen-Nassau)	Germany
PHi.	*Polizeischule Hildesheim* (Police School, Hildesheim)	Germany
PK.	*Polizeischule Kiel* (Police School, Kiel)	Germany
PI.	*Polizei-Institut* (Police Institute)	Germany
PM.	*Polizeischule Münster* (Police School, Münster)	Germany
PMd.	*Polizeischule Hannoverisch-Münden* (Police School, Hannoverian-Münden)	Germany
P.N.	*Polizeischule Niederschlesien* (Police School, Lower Saxony)	Germany

MAUSER MILITARY RIFLE MARKINGS

Text	Meaning	Country
P.O.	*Polizeischule Ostpreussen* (Police School, East Prussia)	Germany
P.P.	*Polizeischule Pommern* (Police School, Pomerania)	Germany
P.R.	*Pionier-Bataillon, Rekruten-depot* (Engineer Batallion, Recruits' Depot)	Germany
P.R.	*Polizei-Reitschule* (Police Equestrian School)	Germany (Bavaria)
PS.	*Polizeischule Sensburg* (Sensburg Police School)	Germany
P.W.	*Polizeischule Westfalen* (Police School, Westphalia)	Germany
Pw.B.	*Polizeiwehr Bayern* (Police Defense Force, Bavaria)	Germany (Bavaria)
PZ	*Rota pornocného zdravotnictva* (Auxiliary medical company)	Czechoslovakia

MAUSER MILITARY RIFLE MARKINGS

Text	Meaning	Country
P.S.	*Polizeischule Sachsen* (Police School, Saxony)	Germany
P.Sch.	*Polizeischule Schleswig-Holstein* (Police School, Schleswig-Holstein)	Germany
PZ	*Rota pornocného zdravotnictva* (Auxiliary medical company)	Czechoslovakia
PERKUN	*Fabryka Motorów "Perkun"* (The "Perkun" Engine Factory)	Poland
Preduzeće 44	Enterprise 44 (A facility in Kragujevac)	Yugoslavia
R.F.V.	*Reichs-Finanz-Verwaltung* (Reich Finance Administration)	Germany
Rh.P.	*Rhein Polizei* (Rhine Police)	Germany
R.FAMAGUE	*Reformado de la Fábrica de Material de Guerra* (Converted by the War Material Factory)	Colombia*

MAUSER MILITARY RIFLE MARKINGS

Text	Meaning	Country
R.M.	Abbr. for *República Mexicana* (Mexican Republic)	Mexico
R.M.G.	*Reserve-Maschinengewehr* (Reserve regiment machinegun company)	Germany
R. R.	*Reserve-Regiment* (Reserve infantry regiment, 1877-1909; after 1909 indicates *Rekrutierung-Regiment*, a reserve regiment recruiting depot)	Germany
R. R.	After 1909 indicates *Rekrutierung-Regiment*, a regular infantry regiment recruiting depot	Germany
República de El Salvador	Republic of El Salvador	El Salvador
República del Paraguay	Republic of Paraguay	Paraguay
República Mexicana	Republic of Mexico	Mexico

MAUSER MILITARY RIFLE MARKINGS

Text	Meaning	Country
República Oriental	Oriental Republic	Uruguay
República de Perú	Republic of Peru	Peru
República Peruana	Peruvian Republic	Peru
S	*Hradni stráž* (Hradčany Castle Honor Guard, Prague)	Czechoslovakia
S	Found on Gew.98 rifles made in the 1920s	Germany
S.	*Schutzpolizei* (Police force)	Germany
𝔖	Indicates a rifle modified to accept the **sS** cartridge	Germany*
S28	Found on the Kar. 98b	Germany
S.A.	*Société anonyme* (Joint-stock company)	Any French-speaking country
S.B.B.	*Schutzpolizei Berlin (berittene)* (Berlin Mounted Police)	Germany
S.B.Ko.	*Schutzpolizei Berlin Kommando* (Berlin Police [Supply & Maintenace Dept.])	Germany

MAUSER MILITARY RIFLE MARKINGS

Text	Meaning	Country
S.B.M.	*Schutzpolizei Berlin (Mitte)* (Berlin Police, Central District)	Germany
S.B.No.	*Schutzpolizei Berlin (Nord)* (Berlin Police, Northern District)	Germany
S.B.O.	*Schutzpolizei Berlin (Ost)* (Berlin Police, Eastern District)	Germany
S.B.S.	*Schutzpolizei Berlin (Süd)* (Berlin Police, Southern District)	Germany
S.B.So.	*Schutzpolizei Berlin (Südost)* (Berlin Police, Southeastern District)	Germany
S.B.W.	*Schutzpolizei Berlin (West)* (Berlin Police, Western District)	Germany
SRP	*Štábní rota Zemského vojenského velitelství v Praze* (Land military HQ in Prague, Staff Company)	Czechoslovakia

MAUSER MILITARY RIFLE MARKINGS

Text	Meaning	Country
SRR	*Pomocná rota Bratislava* (Bratislava Auxiliary Company)	Czechoslovakia
s.R.R.	*Schweres-Reiter-Reserve* (Heavy cavalry reserve regiment)	Germany
SRU	*Štábní rota Zemského vojenského velitelství v Užgorodu* (Land military HQ in Užgorod, Staff Company)	Czechoslovakia
StKG	*Sturmkampfgewehr* (Combat Assault Rifle; a flare-launching K98k)	Germany
Str.	*Streck* (point)	Sweden*
SV	*Pomocná rota vojenského sboru* (Military Corps Auxiliary Company; added in 1939)	Czechoslovakia
svwMB	Post-World War II marking	France*

MAUSER MILITARY RIFLE MARKINGS

Text	Meaning	Country
SZ	*Pomocná rota* (Auxiliary Company Brno; added in 1939)	Czechoslovakia
Serie C.Ch.	*Serie Carabina Chilena* (Chilean Carbine Series)	Chile
Sikte/Rp	Sight [setting]/correction	Sweden*
Skjutning med spetskula	Shooting with pointed [i.e., *Spitzer*] bullet	Sweden*
Spandau	A part of Berlin; site of an arsenal	Germany
ST.-DENIS	St.-Denis, a part of Paris; site of *Société Française des Armes Portatives*	Uruguay*
Steyr-Solothurn Waffen A.G.	Steyr-Solothurn Weapons, Inc.	Austria
Suhl	A city in Germany; site of an arsenal	Germany
T	*Telegrafní prapor* (Signals Battalion)	Czechoslovakia
T.	*Train-Bataillon* (Logistics support battalion)	Germany*
T.	*Telegraf* (Telegraph unit)	Bavaria, Prussia, Saxony, Württemberg

MAUSER MILITARY RIFLE MARKINGS

Text	Meaning	Country
T.B.	*Train-Bataillon, Feldbäckereikolonne)* (Field bakery column)	Germany
T C	*Türkiye Cümhuriyeti* (Republic of Turkey)	Turkey
TDB	*Velitelství těžkých dělostřeleckých brigád* (Heavy Artillery Brigade HQ)	Czechoslovakia
T.F.	*Train-Bataillon, Fuhrparkkolonne)* (Field transport and supply column)	Germany
tgf	98k made at Brno for East Germany	Czechoslovakia
T.L.	*Train-Bataillon, Feldlazarett* (Field hospital support unit)	Germany
T.P.	*Train-Bataillon, Proviantkolonne* (Field provisions support column)	Germany
T.R.Z.	*Tehnički Remont Zavod* (Technical Repair Factory)	Yugoslavia
𝒯.𝑅.𝒵.5	T.R.Z. 5 *Tehnički Remont Zavod* (Technical Repair Factory No. 5)	Yugoslavia

MAUSER MILITARY RIFLE MARKINGS

Text	Meaning	Country
TS	*Škola pro tělesnou výchovu* (Corps Training School)	Czechoslovakia
T.S.	*Train-Bataillon, Sanitätskompanie* (Field medical support company)	Germany
Torped	Boat-tail [bullet]	Sweden*
ÚZ	*Ústřední zbrojnice* (HQ Arsenal; added in 1939)	Czechoslovakia
UZN	*Učiliště ženijní* (Sapper Training College)	Czechoslovakia
V	*Vozatajský prapor* (Mounted Battalion)	Czechoslovakia
VA	*Vojenská akademie* (Military Academy)	Czechoslovakia
VAM	*Válečný archív a museum* (Military Archive and Museum)	Czechoslovakia
VCHU	*Vojenský chemický sklad* (Military Chemical Depot)	Czechoslovakia

MAUSER MILITARY RIFLE MARKINGS

Text	Meaning	Country
VG	*Volkssturm Gewehr* (People's Army Rifle)	Germany
VH	*Vojenský hřebčinec* (Military Stud Farm)	Czechoslovakia
VP	*Vodní prapor* (River Battalion)	Czechoslovakia
VPB	*Vojenská policie Bratislava* (Military Police, Bratislava)	Czechoslovakia
VS	*Vojenská válečná škola* (Military War College)	Czechoslovakia
VT	*Vojenská trestnice* (Military Prison)	Czechoslovakia
VT	*Vojenská tovarná* (Military Factory; added in 1939)	Czechoslovakia
VTLU	*Vojenská technický a letecký ústav* (Military Technical and Aviation Dept.)	Czechoslovakia
VV	*Vojenská věznice* (Military Prison)	Czechoslovakia

MAUSER MILITARY RIFLE MARKINGS

Text	Meaning	Country
VZS	*Vojenský zdravotnický sklad* (Military Medical Depot)	Czechoslovakia
VZU	*Vojenský zeměpisný ústav* (Military Geographical Institute)	Czechoslovakia
V. Chr. Schilling	*Victor Christian Schilling*, a firearms company	Germany
Vz.	*Vzor* (Model)	Czechoslovakia
W	Wilhelm II (1891-1918), king of Württemberg	Württemberg
W (on the receiver)	(Rifle has been test-fired for accuracy)	Poland
Wi.M.	*Wirtschaftsamt München* ([Police] Economic Affairs Office, Munich)	Germany (Bavaria)
Wi.N.	*Wirtschaftsamt Nürnberg* ([Police] Economic Affairs Office, Nuremberg)	Germany (Bavaria)
W.P.	*Wojsko Polskie* (Polish Army)	Poland

MAUSER MILITARY RIFLE MARKINGS

Text	Meaning	Country
Wz.	*Wzór* (Model)	Poland
Wz.29	*Wzór* (Model) 1929	Poland
Wz.29/40 (on the siderail)	Polish Wz.29 carbine converted by the Germans to a K98k facsimile	Poland
Waffenwerke Oberspree, Kornbusch & Co.	Oberspree, Kornbusch & Co. Arms Factory	Germany
WARSZAWA	Warsaw	Poland
Z	*Vojenská zásobárna* (Military warehouse)	Czechoslovakia
Z	Technical inspection mark (1922-1939)	Poland
Z.A.R.	*Zuid Afrikaansche Republiek* (South African Republic)	South Africa*
Zb.	Abbreviation of *Zbrojownia* (Arms factory)	Poland
ŽL	*Železniční pluk* (Railroad Regiment)	Czechoslovakia

63

MAUSER MILITARY RIFLE MARKINGS

Text	Meaning	Country
ZLS	*Zemský letecký sklad* (Provincial Aviation Depot)	Czechoslovakia
Zn	Believed to indicate a rework at the *Reichswehr* facility in Zeithain	Germany
ŽN	*Ženijní pluk* (Engineer Regiment)	Czechoslovakia
ZZ	*Zemská zbrojnice* (Provincial Arsenal)	Czechoslovakia
Zbrojovka Brno A.S. Vz. 24	Brno Arms Factory, *Akciový Společnost* (Joint-stock Co.) *Vzor* (Model) 24	Czechoslovakia
1920	Weimar-era military property mark	Germany
1937	Found on the receiver ring; these vz. 24 rifles have a P-prefixed serial number	China
7,9	The bore diameter; in this case 7.9 mm	Germany*

MAUSER MILITARY RIFLE MARKINGS

Text	Meaning	Country
7.62	Indicates a rifle made for ---or reworked to--- the standard NATO caliber (7.62 x 51 mm)	Israel, Spain, and others
98.Th.	Indicates a reworked rifle with a Model 98 action	Turkey*

NOTES

Belgium ---

French and Flemish (almost identical to Dutch) are the official languages of Belgium. Generally, both languages must be used on government documents, property, etc. The **ABL** inscription is bilingual, with the **B** for *belge* and *Belgisch* ("Belgian", in French and Flemish respectively) overlapping and fitting in nicely with the grammatical structure of the two languages.

Fabrique National, *Fabrique National Herstal-Liège*, *F.N.*, and so forth are all the same facility. Founded in 1889 and located at Herstal-léz-Liège, this factory was one of the most important ---perhaps the most important--- producer of Mauser rifles after the Mauser factory at Oberndorf was degraded due to limitations imposed by the Treaty of Versailles on post-World War I German arms manufacturing.

MAUSER MILITARY RIFLE MARKINGS

Brazil ---

The *Força Policia do Distrito Federal* was a paramilitary police force formed in Rio de Janeiro in ---as far as I have been able to determine--- the 1930s, when Rio was Brazil's capital. When the capital was moved to the brand new city of Brasilia in 1960 the *Força* moved there as well.

Colombia ---

When Colombia switched to .30-'06 ammunition in the early 1950s it converted a number of cal. 7 mm rifles already in stock to the new caliber. This work was done at the Fábrica de Material de Guerra (War Material Factory) in the Colombian capital, Bogotá; these rifles are identified by the marking **.30** on the chamber and the **R.FAMAGUE** marking described in the chart.

Czechoslovakia ---

All of the various **ČS** *Zbrojovka* inscriptions refer to the same factory at Brno, now in the Czech Republic. This facility has undergone numerous name changes during its existence, beginning after World War I in 1919 as the State Armament and Engineering Works. A successor to the arms works ---privatized and prosperous in the aftermath of the collapse of Communism--- is still in business making high quality firearms, and is usually referred to simply as **ČZ** (for *Česká Zbrojovka*, Czech Arms Factory); it is now located inUherský Brod.

MAUSER MILITARY RIFLE MARKINGS

Dutch East Indies ---

For a discussion of the I.O.B. see the **Notes** at the end of the **Emblems** chapter.

Finland ---

The *Suojeluskunta* (Civil Guard) was a sort of Finnish national militia from about 1920 until 1944. This organization's rifles were principally the numerous models of Mosin-Nagants[*], but several types of Mausers were also in common use: mainly the Swedish M94 carbine and M96 infantry rifle, and the German *Gewehr* 98, 98a carbine, and 98k rifle. At one time or another the Civil Guard arsenal also included Arisakas, Berdans, Carcanos, Mannlichers, Simonovs, Tokarevs, Vetterlis, Winchesters, and just about anything else that could be used to fire a harmful projectile at the Russians.

France---

After World War II the French issued captured Mauser rifles to the Foreign Legion, border guards and other units; large numbers of these rifles ended up in Indochina where they were used both by and against the French and, later, against the Americans as well. These French-issued weapons were marked with a star, which was the French nitro proof mark ---see the **Emblems** chapter for an illustration --- and retained the wartime the codes for the

---

[*]For a more detailed treatment of this subject see *The Mosin-Nagant Rifle*, Terence W. Lapin. (North Cape Publications: Tustin, CA, 1998.)

plants. The entry in the foregoing chart shows **svw**, the late World War II code for the Mauser factory at Oberndorf, which was occupied and operated by the French in 1945 and 1946. Some rifles produced under French control were marked with the two-digit date 45, for 1945, but some were also marked with date codes for 1945 and 1946; the letters **MB** are the date code for 1945.

Germany ---

   The Reichs: The term *Deutsches Reich* (literally, German Empire) can be the source of some confusion: it does not apply solely to the 1933-45 Hitler era "Third Reich" (*Drittes Reich*), but to the first two Reichs as well.
   The First Reich was the empire of Charlemagne, who was crowned Emperor of the Romans in 800 A.D. This was a largely a symbolic title as the Roman Empire as such had ceased to exist over 300 years earlier, when the last emperor of the West, Romulus Augustulus, was quietly but forcibly retired to his rural estates in 476 A.D. The Second Reich began in January 1871, when the numerous German states were united under the king of Prussia, Wilhelm I, who then became emperor (in German, *Kaiser*) of the empire as well as king of Prussia. *Kaiser*, by the way, is simply a mutation of "Caesar". The Second Empire lasted until the end of World War I in November 1918, when Kaiser Wilhelm II abdicated and fled to the Netherlands. (There was another Kaiser, Friedrich III, between the two Wilhelms, but he lasted less than a year in 1888 before dying of cancer).

   Military Units: As to general organization I will mention only that the term *Reichswehr* ("Reich Defense") refers to the post-World War I German armed forces as a

MAUSER MILITARY RIFLE MARKINGS

combined command, officially so named as of January 1, 1921; the army as a service branch was called the *Reichsheer* ("Army of the Reich"). In February 1934 the name of the combined command was changed to *Wehrmacht* ("Defense Force").

I am reluctant to get into the organization of the German military of the Mauser Era, especially before the end of the First World War, as it is a broad and complex subject well beyond the scope of this book; however, the proliferation of obsolete European military terms used in the markings has convinced me that a brief glossary would be of value to the reader, thus the following:

Glossary:
Dragoner --- German for "dragoon". The word derives from the French word *dragon*, meaning "dragon", and originally referred to the shape of the hammer of the type of pistols used by such troops; later the word was applied to the pistols, and finally to the troopers themselves. Dragoons were mounted infantry who rode horseback into battle, then dismounted and fought on foot, though sometimes they were used solely as cavalry. At the time of World War I Germany had 28 *Dragoner* regiments.

Garde --- Literally, the French word for "guard". This was a term used for elite troops originally meant to guard the person of monarchs and remains a designation for certain ceremonial and elite units to this day, especially in Great Britain and France.

Grenadier --- As the name implies, this originally meant a grenade-thrower, usually a soldier chosen for his height. Although tall men make bigger targets, military doctrine was that they were useful by virtue of having a greater throwing range, so whole units were formed of them. Some eighteenth-century rulers ---Frederick the Great of Prussia, for example--- thought such men looked very impressive in uniform (Frederick liked men in uniform and, for that matter, men in general), and grenadier regiments gradually evolved into units regarded as elite more for their decorative than their functional attributes. Grenadier regiments were sometimes titled 'Guards' as well. In 1942 the German army revived the unit title *Grenadier* as part of a military moral-building program; oddly enough, in September 1941 the Soviets had done the same thing for the same reason with the title "Guard".

Husar --- In English we add an extra "s" to the word ("Hussar"), but the meaning is the same: a light cavalryman; though when first used in the 15th century the word meant a freebooter. The Germans and Austrians got *Husar* from the Hungarians (*huszár*), who got it from the Serbs (*kursar*), who got it from the Italians (*corsaro*) --- their word for Barbary Coast pirates; in English the word became "corsair". As with other like and similar cavalry types Hussars were often best known for their elegant --- some would say comic-opera--- uniforms.

Kürassier --- Another borrowing from French, in this case *cuirassier*, which referred to the leather (*cuir*, in French) breast-and-back armor worn by light cavalry units. *Kürassiers* were another type of European cavalry with very decorative uniforms. There were ten German *Kürassier* regiments at the time of the First World War.

Landsturm --- In Germany and Austria-Hungary this was a form of third-line militia. Beginning in 1895 German males between the ages of 17 to 20 were subject to conscription into the *Landsturm*, after which they could be required to serve two years in the regular army and five in the reserves (but only three years' reserve duty for cavalry and four for horse artillery). At age 27 a man became liable for *Landwehr* service until age 39, at which point he could be drafted back into the *Landsturm* up to age 45. The

MAUSER MILITARY RIFLE MARKINGS

Landsturm was, theoretically, only to be used within the German borders.

Landwehr --- German reserve units generally employed as support for the regular army; service in the *Landwehr* was compulsory if required. The term *Landwehr* was also used in Austria-Hungary, where the Hungarian equivalent was *honvéd*.

Train --- In this case the word has nothing to do with locomotive engines and freight cars. As in English, the German military usage refers to a force ---or group of elements--- which furnishes logistical support to some other force; it also means a convoy. This word is another borrowing from French that, in or after World War I, was replaced by *Zug*, the usual German word for "train" with its several meanings.

Uhlan --- A light cavalryman or lancer. During the First World War Germany had 24 *Uhlan* regiments. The Germans borrowed the word from the Poles, who had got it from the Mongols. *Uhlan* uniforms had a Polish-influenced look to them, especially the *czapka* --- a distinctive headgear with a hard, flat diamond-shaped top, whose descendant is still used by the Polish military.

Volkssturm --- The word means "People's Assault [Forces]"; it was the World War II German home guard, formed by Hitler's order of September 25, 1944, by which time the country had become desperate for manpower. Theoretically, only men from 16 to 60 were to be drafted into the *Volkssturm* but in practice the scope was greatly widened to encompass the elderly, the adolescent, and the infirm. The *Volkssturm* were armed mostly with captured foreign weapons, obsolete German ones, and several versions of a no-frills Mauser called the VK-98 (*Volkssturm Karabiner-98*, the *Volkssturm* Rifle [Model] 98), which was made in both 1- and 5-shot types.

MAUSER MILITARY RIFLE MARKINGS

Military Unit Markings: I had hoped to avoid this subject, but reader interest convinced me to give a very basic outline.

Military unit-marked weapons in the German imperial forces generally consisted of a number indicating the regiment number, followed by a letter or letters denoting the type of regiment, followed in turn by two sets of numbers: the first indicating the regimental company's number and the last the individual weapon's number. It is easiest to read these designations "backwards": for example, a rifle stamped **2.H.3.44** is rifle No.44 belonging to the 3rd company of the 2nd *Husaren-Regiment* (light cavalry regiment). There are, naturally, exceptions to these general rules.

The German military discontinued the practice of unit-marking their weapons in 1937.

The **BS** marking, stamped into the butt stock, is presumed to mean that the rifle was repaired or reworked at the *Reichswehr* (inter-war German armed forces) depot at Berlin-Spandau. Germany, like many countries, has several types of police, among which are the Railway Police (*Bahnschutz*); this entity used the initials **BS** on rifle receivers; for an illustration see the **Emblems** chapter.

Police Formations and Unit Markings: The subject of German police units and the property markings they used is quite similar in complexity to those of the German military. In an effort to simply the matter I am including a brief description of the types of inter-War police and their marking systems at this point, with tables of regional markings placed towards the end of the book, just before the wartime production codes, although a number of markings do appear in the foregoing chart.

MAUSER MILITARY RIFLE MARKINGS

Einwohnerwehr Bayern, (Bavaria Residents' Defense Force) whose marking is **EWB**, was the Bavarian paramilitary organization of that name, one of a number of such units formed by authority of the Prussian Interior Ministry in April 1919. The chaos afflicting Germany after the end of World War I in November 1918 was so great and so widespread that the German national government was forced to acknowledge the inability of regular police to protect the populace from looters, insurgents, etc.; the solution was to permit localities to form what were in effect their own militias for defense. These groups were considered dangerous by the Allied military occupation authorities, and in late June 1921 the German government ordered *Einwohnerwehr* formations to be disbanded.

Kommunalpolizei (also called *Gemeindepolizei*), both of which mean, approximately, "community police", were the police forces of smaller cities and towns, especially in Prussia.

Kriminalpolizei (the slang term, still in use, is *Kripo*) refers to the detective branch of a police force, usually urban rather than rural.

Landjägerei (also called *Gendarmerie*) were rural constabulary, but the term could also be applied to the police forces of smaller cities and towns.

Schutzpolizei is the generic term for German urban police forces. In and after the police organization reforms of 1922 the *Schutzpolizei* also contained the *Schutzmannschaft* units of pre-World War I imperial Germany and the Sicherheitspolizei (security police) of the immediate post-War years. Both these terms were revived

MAUSER MILITARY RIFLE MARKINGS

during the Nazi era but had quite different meanings, which do not concern us here.

Police Unit Marking System: This was similar to that used by the German military in that the largest unit was listed first, then subunits, and, lastly, the number of the individual piece. The detail involved in property marking (how far down the organization chain the specifics of identification went) varied from one police force or police unit to another, from as simple as **K.B.5** (item #5 of the Berlin *Kriminalpolizei*), to **S.B.S.I 1.7**, for item #7 belonging to Centuria* #1 of the First Command of the South Berlin Police Force.

Readers will notice that some entries will have the German word for "company" spelled *Kompagnie* while others have *Kompanie*. The difference? Not much, really: the first spelling is the older one, borrowed from the French word *compagnie* and 'germanized', which the Germans further germanized after World War I.

Gebrüder Bing (Bing Brothers) was a toy manufacturing firm specializing in metal mechanical toys. The company was located in Nuremberg, once the toy-making capital of the world and still famous for its high-quality toys. During the First World War Bing Bros. made Mauser parts on contract for the German government.

The Germans developed the **sS** (for *schweres Spitzgeschoss*, heavy pointed bullet) after World War I. This was an improved version of the pointed, boat-tail **S** bullet they originated in 1904-05; a new cartridge also was

---

* Some German police forces, like ancient Roman legions, were divided into units of 100 men called a *centuria*.

MAUSER MILITARY RIFLE MARKINGS

developed to accommodate it. Because of its greater weight (198 gr. as opposed to the **S** bullet's 154 gr.) use of the **sS** round required altering rifles' sights to adjust to its different ballistic properties. Pre-existing rifles changed to use the **sS** bullet are so marked.

The G-codes: Foreign Mauser rifles used by the Germans during World War II included many weapons captured or seized from countries they overran. Some types ---generally the older ones or those of less-common calibers--- went to rear-echelon troops, with the more modern ones that used the standard German military 7.92 mm cartridge going to front-line units. For identification and logistical purposes these arms were given code designations, generally including a letter showing the weapon's country of origin: **(b)** for Belgian, **(e)** for English, **(r)** for Russian, and so on; in the case of some types the designations were marked on the arms themselves. I have called them "G-codes" here only because the ones pertinent to our subject start with the letter **G**, for *Gewehr* (rifle), and not because that was the official nomenclature, which was *Fremd-Gerät* ("foreign equipment").

Beginning with their British- and French-approved seizure of the Sudetenland in 1938, and annexation of the rest of Czechoslovakia the following year, the Germans acquired many Czech rifles which, depending on the model, were later designated G. 24(t) for the vz. 24, and G.33/40 for the model 1933 police rifle, adopted by the Germans in 1940. Some M1929 rifles made by the Czechs for ---but, obviously not delivered to--- Colombia were used as the German G. 29/40, and this designation appears also to have been used for the Polish M1929 Mausers seized after the German conquest of Poland in September 1939.

MAUSER MILITARY RIFLE MARKINGS

Bore diameters: The barrels of German-made Mauser rifles produced between 1934 and 1942 are stamped with the actual land-to-land bore diameter in millimeters, being determined by the largest go-plug which would pass through a finished barrel. These numbers will be 7,9 or 7,90, or 7,91, etc. Other bore diameter marking will be found on Israeli and Spanish Mausers rechambered for the standard NATO round.

Orange Free State --- (See **South Africa**)

Poland ---

There are a few unfortunate differences between American firearms terminology and that of some European languages that can easily cause confusion. In Polish *karabin* means "rifle", not "carbine"; the Polish word for "carbine" is *karabinek*. There are also some differences in definitions, which can be summarized by saying that when Europeans refer to "carbines" they usually just mean any shorter version of a similar full-length rifle: not necessarily very light, short-barreled weapons like the U.S. .30 M1 carbine, or short rifles intended for use by cavalry, engineers, etc. Thus, to Germans and Poles, for example, the Gew. 98 is a "rifle" and the Kar98k is a "carbine", though Americans consider both to be "rifles".

Prussia ---

The monogram FW, for *Friedrich Wilhelm*, can be confusing for Mauser purposes as it was also used by Wilhelm I, son of Friedrich Wilhelm IV. In addition to four

MAUSER MILITARY RIFLE MARKINGS

kings of Prussia named Friedrich Wilhelm there was one of Brandenburg who, fortunately, lived well before the Mauser era. By far the most famous king of Prussia was Friedrich II, known to us as Frederick the Great (1712-86; reigned 1740-86). My conjecture is that Wilhelm I assumed the additional "Friedrich" for official purposes as a way to acquire some needed, even if second-hand, prestige, much as David, Prince of Wales, became Edward VIII when his father King George V died, even though David was never crowned. When *his* brother Albert, Duke of York --- "Bertie", to friends and family--- ascended the British throne after Edward VIII's abdication in 1936 he became George VI. In any event the *FW* monogram found on German Mausers of the Imperial period is that of Wilhelm I, King of Prussia and, after January 1871, Emperor of Germany as well.

South Africa ---

This was one of the short-lived republics of the Boers (*boer* is the Dutch and Afrikaans word for "farmer") in what is now South Africa. "Orange" is a reference to the House of Orange, then and now the ruling family of the Netherlands; for obvious reasons orange is also the Dutch national color. Most---though certainly not all--- Boers are of Dutch origin, and their language, Afrikaans, is derived from late 17th century Dutch. Boers fleeing British encroachment founded the *Oranje Vrei Staat* in 1837. After much travail and warfare the British formally annexed the O.V.S. as the "Orange River Colony" in late May 1900.

The shifting compositions, status, and names of the Boer republics can be complicated. For our purposes it is sufficient to note that, as part of their efforts to resist the British the Boers formed a union of the Orange Free State

MAUSER MILITARY RIFLE MARKINGS

and another Boer republic, the Transvaal, in 1896. This federation was the Z.A.R., *Zuid Afrikaansche Republiek*, (South African Republic), and it was this entity that was the losing side in the Boer War of 1898-1900.

D.W.M. sold a large number of Mausers to the Boers during their war against the British (this had German government approval and support as Germany was both sympathetic to the Boer cause and enjoyed the opportunity to nettle the British). Many of these Mausers had "Mod. Mauser" stamped on the receiver; to this day these rifles are called "Mod Mausers" in South Africa.

Spain ---

The inscription *Generalitat de Catalunya Comissió de L'Industria de Guerra Subsecretaria de Armament*, though appearing on some Spanish M1916 carbines, is not Spanish: it is Catalan, the language of Catalonia, a region in northeastern Spain. Historically, the term *Generalitat de Catalunya* has been used to refer to the autonomous government of Catalonia and the city of Valencia. There have long been strong regional desires in this region for greater autonomy or political independence from Spain, and these strivings were partly the cause of particular violence and suffering there during the Spanish Civil War.

Sweden ---

The markings *Överslag*, *Torped* and *Str.* are found on the metal discs inset in the butt stock of some Swedish Mausers. They refer to the adjustments a shooter must make to accommodate firing the M94/41 cartridge in rifles originally sighted for the older M94 round. The metal disc method was replaced by a decal giving similar ranging information, which is why not all Swedish Mausers will be found with the disc.

The markings *Avstånd, Sikte för trubbkula, Sikte Rp,* and *Skjutning med spetskula* are found on the butt stock decals giving firing information which replaced the stock discs mentioned above.

Turkey ---

AS.FA and *Kırıkkale Tüfek Fabrikası* (Kırıkkale Rifle Factory) were predecessors to *Makina ve Kimya Endüstrisi Kurumu* (Machine and Chemical Industry Co.), perhaps better known to firearms enthusiasts as the ammunition manufacturer MKE.

There is some question about the exact meaning of **AS.FA**. *As.* is a common Turkish abbreviation for *askerî*, meaning "military", and it has been suggested that AS.FA is an abbreviation of something yet undetermined. Until a definitive answer is found I will continue with the interpretation found in the chart, which was that conveyed to me by the Turkish military when I inquired about it.

The **Th.** part of the marking **98.Th.**, found on some late 1940s-early 1950s Turkish Mauser rifles, may stand for *tahvil*, meaning "conversion" or "change". I suggest that this marking indicates rifles converted from earlier Turkish Mauser models (the M1893, for example) to the M1898. I cannot say with any degree of certainty why this marking did not appear on earlier reworks of 7.65 x 53 mm and other older rifles to the more modern 7.92 x 57 mm caliber, which Turkey began in volume around 1935.

MAUSER MILITARY RIFLE MARKINGS

Uruguay ---

Why does a South American country call itself an "Oriental" republic? Because it is located on the eastern (in Spanish, *oriental*) banks of the Uruguay River and the Río de la Plata.

At the end of the 1890s Uruguay ordered some strange hybrid weapons from Europe: M1871 Mauser rifles and carbines converted to fire the 6.5 x 53.5 mm Daudetau cartridge. The conversion work was done by *Société Française des Armes Portatives à St.-Denis* (French Small Arms Co. at St.-Denis), whose plant was located in St.-Denis, a district of Paris. The words ST.-DENIS are stamped into the barrels of these weapons. St.-Denis (in English, "St. Dennis"), by the way, is the patron saint of France.

Yugoslavia ---

The principal Yugoslav arms factory during the "Mauser Era" was the Military Technical Factory, or *Voini Tekhnichki Zavod* (in Serbian, Војни Технички Завод), which used as its mark the letters ВТЗ (i.e., VTZ). The plant was located at Kragujevac, in the Yugoslav republic of Serbia. See *Kragujevac* in the **Gazetteer** for further details.

Some Mausers have been refurbished at a generic repair and overhaul facility near Sarajevo, Technical Repair Establishment No. 5 (in Croatian, *Tehnički Remont Zavod 5*), whose marking is **T.R.Z. 5**; this marking appears on items both in block letters and in script.

Zuid Afrikaansche Republiek --- (See **South Africa**)

MAUSER MILITARY RIFLE MARKINGS

INSCRIPTIONS IN NON-WESTERN LANGUAGES

Because of the variety of languages and writing systems involved the inscriptions in this section are listed in alphabetical order by country. [Note: "Abbr." = abbreviation.]

WRITING	MEANING	COUNTRY
武年一十二 廠製器兵東廣 造年一十二	Year 21 Model Made by Kwantung Arsenal Made in the Year 21 [i.e., 1932]	China
中正式	China Standard Model (Kung-hsien Arsenal)	China
中正式	China Standard Model (Hanyang Arsenal)	China

MAUSER MILITARY RIFLE MARKINGS

WRITING	MEANING	COUNTRY
	China Standard Model (Arsenal #21)	China
	Han Model (Arsenal #21)	China
	Model 46 (Kung-hsien Arsenal)	China
	Zhejiang Iron Works	China
	Model 77	China
	10,000 [Abbr. of] *"His Imperial Majesty Haile Selassie I, King of Ethiopia"*	Ethiopia

MAUSER MILITARY RIFLE MARKINGS

WRITING	MEANING	COUNTRY
	In the center: *His Imperial Majesty Haile Selassi I, King of Ethiopia* To the sides of the lion's head: *1965 year*	Ethiopia
ΒΑΣΙΛΕΙΟΝ ΤΗΣ ΕΛΛΑΔΟΣ	Kingdom of Greece (lit., "Kingdom of the Greeks")	Greece
ΥΠ 1930	Abbr. for **ΥΠΟΔΕΙΓΜΑ** (*ipodheigma*) **1930** (Model 1930)	Greece
צבא ההגנה לישראל	*Tsvā' haHagānā l'Yisraēl* (Israel Defense Force)	Israel
پیاده	*Piyādeh* (Infantry)	Persia

MAUSER MILITARY RIFLE MARKINGS

WRITING	MEANING	COUNTRY
تفنگ نمونه ۱۳۰۹ کارخانه اسلحه سازی برنو Note: The actual inscription is in one line, but for clarity it is shown here in 2 lines. The 2nd line begins with the word *Kārkhāneh* ("factory") at the right.	*Tofang-e namunah 1309 Kārkhāneh eslihah-e sāzi Brno* (Rifle Model 1309 Brno Arms Factory)	Persia
تفنگ بلند نمونه ۱۳۱۷ کارخانه اسلحه سازی برنو Note: The actual inscription is in one line, but for clarity it is shown here in 2 lines. The 2nd line begins with the word *Kārkhāneh* ("factory") at the right.	*Tofang-e boland namunah 1317 Kārkhāneh eslihah-e sāzī Brno* (Long Rifle Model 1317, Brno Arms Factory	Persia

MAUSER MILITARY RIFLE MARKINGS

WRITING	MEANING	COUNTRY
تفنگ متوسط نمونه ۱۳۱۷ کارخانه اسلحه سازی برنو Note: The actual inscription is in one line, but for clarity it is shown here in 2 lines. The 2nd line begins with the word *Kārkhāneh* ("factory") at the far right.	*Tofang-e-motavasset namunah 1317 Kārkhāneh eslihah-e sāzi Brno* (Medium Rifle Model 1317 Brno Arms Factory)	Persia
کارابین نمونه ۱۳۲۸ ساخت اسلحه سازی ارتش Note: The actual inscription is in one line, but for clarity it is shown here in 2 lines. The 2nd part begins with the word *sākht* at the far right.	*Kārabīn namunah 1328 sākht eslihah-e sāzī artesh* (Model 1328 Carbine made by the Army Arsenal)	Persia
๑๒๑ ร.ศ.	R. S. 121 (Bangkok Era 121) [1902 A.D.]	Siam (Thailand)

85

MAUSER MILITARY RIFLE MARKINGS

WRITING	MEANING	COUNTRY
๑๖/๒๗ ๔ บ บ น	Model 47/66	Siam (Thailand)
الإدارة الشرطي	*Ad-Darak as-Sūrī* (The Syrian Police)	Syria
رسومات	*Rüsumat* (Customs Administration)	Turkey
سنة	*Sene* (Year)	Turkey
متره	Meter	Turkey
المنياده نقار حلنده اوبرن دورفده / ماوزر سلاح فابريقه سى Note: Because of its length the inscription it is shown here divided into two parts. The 2nd line begins with the word *Mauser*.	*Almanya'da Nekar Sahilinde Oberndorf'da Mauser Silah Fabrikası* (Mauser Arms Factory, Oberndorf am Nekar, Germany)	Turkey

MAUSER MILITARY RIFLE MARKINGS

WRITING	MEANING	COUNTRY
اوبرندورفده ماوزر سلاح فابريقه سى	*Oberndorf'de Mauser Silah Fabrikası* (Mauser Arms Factory, Oberndorf)	Turkey
اوبرنو ورفذه ماوزر سلاح فابريقه سى	*Oberndorf'de Mauser Silah Fabrikası* (Mauser Arms Factory, Oberndorf)	Turkey
АРТ.ТЕХ. ЗАВОД - КРАГУЈЕВАЦ	*Art. Tekh. Zavod – Kragujevats* (Artillery-Technical Factory, Kragujevac)	Yugoslavia
АУСТРИЈСКА ОРУЖНА ФАБРИКА ШТАЈЕР	*Austriyska Oruzhna Fabrika Shtayer* (Austrian Arms Factory, Steyr)	Yugoslavia

MAUSER MILITARY RIFLE MARKINGS

WRITING	MEANING	COUNTRY
ВОЈНОТЕХ. ЗАВОД - КРАГУЈЕВАЦ	*Voynotekh. Zavod Kragujevats* (Military-Technical Factory, Kragujevac	Yugoslavia
КРАЉЕВИНА ЈУГОСЛАВИА	*Kralyevina Yugoslavia* (Kingdom of Yugoslavia)	Yugoslavia
МОДЕЛ 1924	Model 1924	Yugoslavia
МОДЕЛ 1924 ЧК	Model 1924 ChK (Indicates rifle used by the Chetniks)	Yugoslavia
ОРУЖНА ФАБРИКА МАУЗЕРА А.Д. ОБЕРНДОРФ А. Н.	*Oruzhna Fabrika Mausera A.D. Oberndorf A/N* (Mauser Arms Factory, Inc. Oberndorf A/N)	Yugoslavia
ПОЛИЦИЈА НИШ	*Politsia Nish* (Nish [a city] Police)	Yugoslavia
ПРЕДУЗЕЋЕ 44	*Preduzetche* 44 (Enterprise [i.e., Industrial Plant] No. 44)	Yugoslavia

MAUSER MILITARY RIFLE MARKINGS

WRITING	MEANING	COUNTRY
РАДИОНИЦА 145	*Radionitsa 145* (Workshop 145)	Yugoslavia
ФНРЈ	FNRJ (Abbreviation for *Federativna Narodnaya Republika Jugoslavia* [Federative People's Republic of Yugoslavia])	Yugoslavia

NOTES

China ---

Han is the revered original ancestor of the ethnic Chinese, who sometimes call themselves the "sons of Han". "Han Model" is thus synonymous with "Chinese Model". Both of these logos depicted in the chart include a swastika[*], which was the mark of China's Arsenal No. 21; this facility was part of the arms manufacturing complex at Hanyang.

There is an element of ambiguity as to the exact meaning of the inscriptions whose translations I have given as "Han Model" and "China Standard Model"; this is understandable given the inherently ambiguous nature of Chinese writing; in written Chinese context is crucial.

---

[*] See the **Notes** on China in the **Emblems** chapter.

MAUSER MILITARY RIFLE MARKINGS

A common Chinese Mauser of the 1930s is often referred to as the "Chiang Kai-shek Model", after the Chinese leader of the time. In Chinese, *Chiang Kai-shek* is written 蔣中正. The ideograph on the left is the family name "Chiang"; the center- and right ideographs are written exactly the same as for "China (or Chinese) standard", as well as forming part of the late statesman's name. Chiang Kai-shek was also known as 蔣介石 (Chiang Jye-shr), and I suggest that if the rifle were named for him it would have used his surname *Chiang* (in Chinese the family name comes first), not his personal names *Kai-shek*. No one, for example, refers to the founder of Communist China, Chiang's mortal enemy, as "Tse-tung", but always as "Mao".

Along the same line of reasoning, the complete written form of the city Hanyang is 漢陽, but only the first ideograph, 漢, the same ideograph that means both China itself (metaphorically) and *Han*, the founder of the Chinese race, appears in the inscriptions. Sometimes in colloquial Chinese only one element of compound place-names is used (for example, *Beijing*, literally, "Northern Capital", is occasionally called simply "*Jing*", "Capital"), but if we consider the rendering of the second inscription in the chart as "China [Standard] Model", then, as I mentioned above, the synonymous "Han Model" is a more logical attribution than "Hanyang Model". I am merely suggesting these interpretations; I do not claim they are definitive. I leave that task to scholars of Chinese.

I will mention at this point that most of the calligraphy in the Chinese logos shown in the charts is stylized, even as various fonts and styles, such as italics and boldface, are used for Roman letters in Western languages. The usual, i.e., non-stylized, Chinese characters for "Chinese Standard Model" are 中正式, and for "Han Model" they are 漢式. The Model 46 marking is stylized as well, and lest it confuse

MAUSER MILITARY RIFLE MARKINGS

anyone looking at the Chinese numerals table in the **NUMBERS** chapter the center figure is merely a version of the more common form of the number 6, 六, and is not a different ideograph.

  Kwantung is a region in northeastern China, to all intents and purposes a large part of Manchuria. The area is now called *Liaoning* Province. The provincial capital city is *Shenyang*, formerly called *Mukden* and site of the arsenal of the same name.

  The ideograph 瓦 is a stylized version of 民, literally "people", which in this instance is an abbreviated form of 民兵, meaning "militia" or "militiaman"; it is found on some Chinese Mausers, usually as a large character on the butt stock. The two ideographs for "militia" also appear in their own right on certain rifles.

  The "Model 77" referred to in the Chart is a locally-produced Chinese Mauser-type rifle, very similar to the Chinese Standard (or "Chiang Kai-Shek") model. "Model 77" refers to July 7th (the seventh day of the seventh month) of 1937, the date of the "Marco Polo Bridge Incident", which the Chinese regarded as the beginning of their war with Japan.

  The top half of the Model 77 marking is comprised of two stylized Chinese sevens 七七 over a stylized version of the character 式 meaning "model" or "type". This rifle was made originally at the Zhejiang Iron Works in eastern China and subsequently at three subunits (Xiao Siun, Shi Tou, and Yu Si). During the course of World War II the factory was moved to Nanping to avoid Japanese occupation. The Model 77 was produced from 1938 until 1946.**

** For more on this rifle see the excellent article by Stan Zielinski and Bin Shih noted in the Bibiliography.

MAUSER MILITARY RIFLE MARKINGS

On some Chinese Mausers a short series of numerals and a symbol appear on the receiver ring: a one-, two- or three-digit number, followed by an arsenal logo, then another one- or two-digit number. The number before the logo is the year in which the rifle was made, and the one after the logo gives the month of the year. In the example at the right the markings show a rifle made in the Year 21 at Arsenal No. 21 (the swastika logo) in the 5th month of the year; a rifle bearing this set of marks would have been made at Hanyang in May of 1932. The custom of indicating on a weapon the month, as well as the year, it was produced was retained in China until at least the mid-1950s.

Ethiopia ---

The writing shown in the first Ethiopian entry in the chart, found on some Ethiopian Mausers, was meant to be a facsimile of Haile Selassie's own handwritten signature. The meaning of the *10,000* is obscure. It has been theorized that rifles thus marked were issued to the late Emperor's bodyguard.

Haile Selassie I was deposed in a Marxist military coup d'état in 1974. On August 28, 1975, a little less than a year later, the new régime announced that he had "died in his sleep". People who should know say that on the night of August 25th two Ethiopian army lieutenants helped the Emperor out of this place of wrath and tears by holding a pillow over his face. Perhaps he should have armed his bodyguards better.

The man who assumed the throne name *Haile Selassie* ("Might of the [Christian] Trinity") was born Prince Tafari Makonnen. In Amharic "Prince Tafari" is *Ras Tafari*. History's loom is large and the patterns woven on it are complex and often strange: Ras Tafari's memory lives on with his devotees, the Rastafarians, for whom he is a demigod.

Shortly after the first edition of this book was published the Emperor's remains were exhumed from their original burial place (the palace latrine), and were interred in the cathedral at Addis Ababa with the proper ceremonies and rites of the Ethiopian Orthodox Church.

Siam (Thailand) ---

For an explanation of the Bangkok Era system see the **Calendars and Dating** chapter.

Yugoslavia ---

All Yugoslav inscriptions in the chart are in Serbian, i.e., written in the Serbian version of the Cyrillic alphabet (for a description of the properties of Serbian and Croatian see the **Alphabets** and **Languages** chapters.) Croatian inscriptions, being in the Roman alphabet, are found in the **Inscriptions in Western Languages** section.

The *Chetniks* were guards for the Yugoslav monarchy between the World Wars. During the Second World War anti-communist Yugoslav monarchist partisans used the name.

MAUSER MILITARY RIFLE MARKINGS

COATS OF ARMS

I will not go into the technical differences between "coat of arms", "shield", and other heraldic nomenclature; the fascinating art and science of heraldry require ---and have--- volumes devoted solely to them. For simplicity's sake they are all referred to here as "coats of arms" or just "arms".

The following table is mostly self-explanatory, but I ask the reader to note that the column titled "Country" will, in some instances, include an organization, (e.g., a military organization) or even a commercial company. Unless otherwise specified the coat of arms is the national one. Countries and organizations can and do change their insignia from time to time, and the reader should be aware that there are variations among some of the coats of arms which may appear on Mausers but that might not be shown below; in all instances they are similar enough to be identified from the following illustrations. For additional trademarks and other entries see the **Emblems** and **Inscriptions in Non-Western Languages** chapters.

As in some of the other sections of this book I have added notes at the end of the present section in which are discussed various aspects of the items illustrated in the charts.

MAUSER MILITARY RIFLE MARKINGS

COAT OF ARMS	COUNTRY
	Argentina
	Argentina (Military Academy insignia)
	Argentina (Navy insignia)
	Argentina (Naval Academy insignia)
	Argentina (Insignia of the University Battalion of the Federal Capital)

MAUSER MILITARY RIFLE MARKINGS

COAT OF ARMS	COUNTRY
	Belgium
	Belgium (Monogram of King Leopold III, reigned 1934-1951)
	Belgium (Monogram of King Baudouin, reigned 1951-1993)
	Belgium (Bayard Co. trademark)
	Belgium (Fabrique National trademark)

MAUSER MILITARY RIFLE MARKINGS

COAT OF ARMS	COUNTRY
	Brazil
	Chile
	Colombia
	Costa Rica

MAUSER MILITARY RIFLE MARKINGS

COAT OF ARMS	COUNTRY
	Czechoslovakia
	Dominican Republic
	Ecuador
	Ethiopia (Below the crown is the monogram of Haile Selassie I)

MAUSER MILITARY RIFLE MARKINGS

COAT OF ARMS	COUNTRY
	Ethiopia (Imperial guard?) For a translation see **NON-WESTERN INSCRIPTIONS**
	Germany (2nd Reich, 1871-1918)
	Germany (3rd Reich, 1933-1945)
	Germany (3rd Reich, 1933-1945); *National Sozialistiches Kraftfahrkorps* (Nazi Motor Corps emblem)

MAUSER MILITARY RIFLE MARKINGS

COAT OF ARMS	COUNTRY
	Germany (Trademark of DWM: Deutsche Munitions- und Waffen Fabrik)
	Germany (Trademark of Ludwig Loewe & Co.)
	Germany (Mauser Co. trademark until 1909)
	Germany (Mauser Co. trademark after 1909)
	Greece

MAUSER MILITARY RIFLE MARKINGS

COAT OF ARMS	COUNTRY
	Guatemala
	Iraq (until the 1958 revolution)
	Israel (Israel Defense Force)
	Mexico

MAUSER MILITARY RIFLE MARKINGS

COAT OF ARMS	COUNTRY
	Netherlands (Monogram of Queen Wilhelmina, reigned 1890-1948)
	Netherlands (Monogram of Queen Juliana, reigned 1948-80)
	Nicaragua
	Paraguay

MAUSER MILITARY RIFLE MARKINGS

COAT OF ARMS	COUNTRY
	Persia (Iran)
	Peru
	Poland
	Portugal (Monogram of King Carlos I, reigned 1889-1908)

MAUSER MILITARY RIFLE MARKINGS

COAT OF ARMS	COUNTRY
	Portugal
	Prussia
	Romania (Crest of King Carol II, reigned 1930-1940)
	Romania (Crest of King Michael I, reigned 1927-30 and 1940-1947)

MAUSER MILITARY RIFLE MARKINGS

COAT OF ARMS	COUNTRY
	Serbia
	Slovakia
	Spain (until 1931)
	Spain (1938-)

MAUSER MILITARY RIFLE MARKINGS

COAT OF ARMS	COUNTRY
	Spain (Air Force)
	Spain (*Guardia Civil*, the National Gendarmerie)
	Spain (La Coruña arsenal)
	Sweden (Carl Gustafs Factory trademark)
	Syria

MAUSER MILITARY RIFLE MARKINGS

COAT OF ARMS	COUNTRY
	Turkey
	Uruguay
	Venezuela
	Yugoslavia (Kingdom, 1921-1945)

MAUSER MILITARY RIFLE MARKINGS

COAT OF ARMS	COUNTRY
	Yugoslavia (Monogram of King Alexander I, reigned 1921-34)
	Yugoslavia (Communist era, 1945-91)

NOTES

Belgium ---

The *Fabrique National* logo is one of the most famous trademarks in the armaments world. The *Bayard* company, although today known largely by its participation in manufacturing the Bergmann-Bayard pistol, also produced Mauser rifles for the Germans during their occupation of Belgium during World War I.

Germany ---

Ludwig Loewe (1837-1886) was a German industrialist with many successful businesses, including the manufacture of sewing machines and coin-minting presses as well as firearms. In addition to running its own arms factory in Berlin, Ludwig Loewe & Co. bought out the Mauser brothers in December 1887, though the name

MAUSER MILITARY RIFLE MARKINGS

"Mauser" was retained. After Ludwig's death his brother, Isidore, controlled the company.

The Loewe Company acquired a controlling interest in Fabrique National, and expanded its firearms business yet again in November 1896, when it merged with Deutsche Metallpatronenfabrik, the Rheinische-Westfällische Powder Company and the Rottweil-Hamburg Powder Co. to form Deutsche Waffen-und Munitionsfabriken (D.W.M.). Reorganization after Germany's defeat in World War II converted D.W.M. to I.W.K. (Industrie-Werke Karlsruhe); this is still the Mauser parent firm.

Israel ---

The shield shown in the chart is not, as some publications have captioned it, the arms of the State of Israel; it is the symbol of the Israel Defense Force and actually has the IDF name on it. See the **Inscriptions in Non-Western Languages** chapter for more about this.

Persia ---

The crest on Persian Mauser receiver rings is the astrological device of the Sun in Leo: a lion ---in this instance holding a scimitar--- with a rising sun behind him. This is a very old symbol going back at least to the Seljuk Turk rulers of Persia during the 13th century, and probably represented the horoscopic birth sign of Kai-Khusrū II, an important ruler of the era. This crest is surrounded by a wreath of laurel leaves and oak leaves and is surmounted by a Persian imperial crown. To no one's surprise this emblem was discarded after the Iranian revolution of 1979.

MAUSER MILITARY RIFLE MARKINGS

Slovakia ---

This coat of arms, showing a patriarchal cross on symbolic representations of the country's Fatra, Matra and Tatra mountain ranges, is found on Mausers used by the Nazi puppet state of Slovakia. Slovakia did not exist as a state before 1939; this emblem was lifted from a 19th century Slovak patriotic organization. The symbol was resurrected as the emblem of the Slovak Republic when it split away from Czechoslovakia in 1993. A close look at the Czechoslovak lion will reveal this emblem on his breast.

Spain ---

The 1931-38 gap between Spanish coats of arms reflects the fact that King Alfonso XIII abdicated in 1931; the symbols of the House of Borbón, including the royal arms, went into exile with him. A republic was declared, which passed through one right-wing and several increasingly bloody left-wing phases, culminating in the Civil War (1936-39). After the war the former symbols of Spain, including a somewhat royalist eagle, were brought back. Note the old royal arms on the eagle in the La Coruña arsenal marking.

The fasces-and-sword emblem found on some Spanish Mausers, is not, as so often stated, the *Falange* Party emblem (that is an ox yoke with five arrows); it is, rather, the emblem of the *Guardia Civil*, the national gendarmerie, and has been since the 1840s. The *Falange* Party ---not quite Fascist, but close to it--- was not Generalissimo Francisco Franco's party: that was the

National Party. Following the Nationalist victory in the Spanish Civil War the *Falange* was gradually absorbed into it over a period of years.

The fasces (from the Latin word *fasces*, meaning a bundle [of sticks]) is in fact an ancient Roman symbol. Fasces were carried by the *lictor*, a sort of court bailiff who attended the highest officials such as consuls, proconsuls and so on. The sticks in the bundle represented the official's authority to punish by flogging, and the axe --- his authority to execute wrongdoers.* Mussolini's Fascist Party adopted the symbol as its own (hence the name), but it has been used in a number of places over the centuries: look at the reverse of a U. S. "Mercury" dime, for example. People interested in military affairs have probably come across the word "fascine" ---it exists in a number of languages--- for bundles of sticks used in constructing trenches, bridging ditches, etc.; it has exactly the same root as *fasces*.

Syria ---

The Syrian falcon had three stars on its breast shield until 1958, when Syria joined Egypt in the short-lived United Arab Republic; at that time the falcon lost a star and got its wings folded. After the breakup of the UAR in 1962

---

*Under Roman law it was illegal for Roman citizens to be flogged without a trial and conviction; this was a useful privilege of Roman citizenship. For an example of this right being invoked see the biblical book of *The Acts of the Apostles*, Chapter 22, verses 25-26, where St. Paul (a Roman citizen) stops a centurion, who is about to flog him illegally, from getting into serious trouble. Another benefit of citizenship was that Roman citizens could not be crucified: they were entitled to claim the right to be beheaded for capital ---from the Latin *capitalis*, literally "of the head (*caput*)"--- crimes, thus the axe in the *fasces*.

MAUSER MILITARY RIFLE MARKINGS

Syria reverted to its former falcon, but with two stars on the shield. The illustration in the chart depicts the Syrian emblem with three stars as found on Syria's Mausers. The bird, incidentally, is indeed a falcon, not an eagle: the falcon was the emblem of the Qureish, the Prophet Muhammad's clan, thus still a powerful symbol in the Muslim world, and turns up periodically in the national emblems of Egypt, Libya, the Palestinian Authority, and others. Some modern Arab sources refer to the bird as the "golden eagle of Saladin" (A.D. 1137?-1193), claiming that it is taken from the personal emblem of the Kurdish leader best known in the West as the opponent of Richard the Lionheart during the Third Crusade.

Turkey ---

For a discussion of the Turkish crescent see the **Emblems** chapter.

Yugoslavia ---

The coat of arms of pre-communist Yugoslavia was adopted in 1921, three years after the birth of the "Kingdom of Serbs, Croats and Slovenes"; this unwieldy name was changed to Yugoslavia, "Land of the Southern Slavs", in 1929. The coat of arms consists of the arms of Serbia (the most prominent), Croatia, and other constituencies, and thus can easily be confused with that of Serbia.

The **29 XI 1943** on the communist Yugoslav coat of arms refers to November 29, 1943, when the Anti-Fascist National Council for the Liberation of Yugoslavia (in effect, Tito's own communist party) established a new national parliament and designated Tito "Marshall of Yugoslavia". In fact, the monarchy was not officially abolished and the communist state declared until late 1945, after the end of World War II. The coat of arms was adopted about a year later, and the date placed on it seems to backdate the new regime by about three years, so giving it an air of established solidity.

MAUSER MILITARY RIFLE MARKINGS

EMBLEMS

The following charts show emblems or symbols found on various Mausers; other illustrations are to be found in the **Coats of Arms** and **Inscriptions in Non-Western Languages** chapters. For a discussion of the meaning of these markings see the **Notes** at the end of this chapter.

Symbol	Meaning	Country
	A *fleur-de-lys*	Bosnia-Herzegovina
	Symbol of the *Guomindang* (Nationalist Party)	China
	Logo of Arsenal #21 at Hanyang	China
	Kung-hsien Arsenal logo	China
	Nanking arsenal logo	China
	Army Ordnance Dept. symbol	China

MAUSER MILITARY RIFLE MARKINGS

Symbol	Meaning	Country
	Model 77 on top of the Zhejiang Iron Works logo	China
	Československá Zbrojovka (Czech Arms Factory) trademark	Czechoslovakia
	Czechoslovakian Army acceptance mark, 1918-22	Czechoslovakia
	Czechoslovakian Army acceptance mark, 1922	Czechoslovakia
E 23	Czechoslovakian Army acceptance mark, 1923-24 (here, for the year 1923)	Czechoslovakia
E 25	Czechoslovakian Army acceptance mark, 1924-37 (here, for the year 1925)	Czechoslovakia
	Indische Ondernemersbond (Indian Entrepreneurs' Federation)	Dutch East Indies

MAUSER MILITARY RIFLE MARKINGS

Symbol	Meaning	Country
(lion image)	Conquering Lion of Judah (Acceptance mark)	Ethiopia
(SA)	*Suomen Armeija* (Finnish Army) property mark	Finland
☆	French nitro proof; as a post-World War II mark it indicates a reissued foreign weapon	France
𝔅 (crowned)	Typical military inspector's mark (prior to the end of World War I)	Germany
M (crowned)	Imperial Navy acceptance mark (M= *Marine* = navy)	Germany
✡	Indicates a *Stern* rifle (see Notes below)	Germany
✶	Indicates a *Stern* rifle (see **Notes** below)	Germany

MAUSER MILITARY RIFLE MARKINGS

Symbol	Meaning	Country
(eagle)	Typical Weimar-era *Beschussstempel* (firing proof)	Germany
M (eagle above)	Navy acceptance mark (M= *Marine* = navy); Weimar era (1923-1933)	Germany
M (anchor below)	Navy acceptance mark Weimar era	Germany
S/91	Pre-1934 acceptance mark	Germany
P/74	Pre-1934 acceptance mark	Germany
D/74	Pre-1934 acceptance mark	Germany
(eagle with swastika)	Nazi-era *Beschussstempel* (firing proof)	Germany
N (eagle above)	World War II-era nitro proof	Germany

117

MAUSER MILITARY RIFLE MARKINGS

Symbol	Meaning	Country
(eagle over L)	Police acceptance mark (C, F and possibly other letters were used as well)	Germany
(eagle over 135)	A typical *Waffenamt* inspector's mark (in this case from the Mauser factory)	Germany
Ⓐ	Parts mark, Berlin-Borsigwalde Mauser factory	Germany
ce	**ce**, the factory code for J. P. Sauer & Sohn	Germany
(eco in circle)	Gustav Genschow & Co AG (Geco) logo	Germany
⚡⚡	Insignia of the *Waffen-SS* (Nazi armed elite guard)	Germany
⚡⚡ZZA4	*SS Zentral Zeugamt No. 4* (SS Central Matériel Office #4; found on some World War II reworks, mostly from Steyr-Daimler-Puch)	Germany

MAUSER MILITARY RIFLE MARKINGS

Symbol	Meaning	Country
⚡	Indicates rifle rework done under *SS* supervision	Germany
☠	*Totenkopf* (Death's Head) with *Sig-Runen*; an *SS* property mark	Germany
△	The Arabic letter *jeem*. Abbr. for *jaysh* (army) [?]	Iraq
△	A variant of the above mark, with a stylized *jeem*	Iraq
✡	I.D.F. inspection mark (stylized Star of David with the Hebrew letter *nūn*)	Israel
⊗	I.D.F. property mark (the Hebrew letter *tsadi*, abbr. for Israel Defense Force)	Israel

MAUSER MILITARY RIFLE MARKINGS

Symbol	Meaning	Country
	"Columns of the Gediminas", the Lithuanian national symbol	Lithuania
	Mukden Arsenal logo	Manchuria
	The Shah's imperial crown	Persia (Iran)
	Fabryka Broni logo	Poland
	Military acceptance mark (1922-1939)	Poland
	Military acceptance mark (1922-1939)	Poland
	Military acceptance mark (1922-1939)	Poland

MAUSER MILITARY RIFLE MARKINGS

Symbol	Meaning	Country
F 2 (in oval)	Military acceptance mark (1922-1939)	Poland
G 2 (in oval)	Military acceptance mark (1922-1939)	Poland
triangle symbol	Technical inspection mark (1922-1939)	Poland
chevron symbol	Technical inspection mark (1922-1939)	Poland
diamond symbol	Technical inspection mark (1922-1939)	Poland
lightning bolt in oval	Technical inspection mark (1922-1939)	Poland
diamond with dot	Technical inspection mark (1922-1939)	Poland

MAUSER MILITARY RIFLE MARKINGS

Symbol	Meaning	Country
⊙ (ni)	Technical inspection mark (1922-1939)	Poland
)(	Technical inspection mark (1922-1939)	Poland
☂	Technical inspection mark (1922-1939)	Poland
♉	Technical inspection mark (1922-1939)	Poland
⬟ (fish)	Technical inspection mark (1922-1939)	Poland
♉	Technical inspection mark (1922-1939)	Poland
◇	Technical inspection mark (1922-1939)	Poland
♡	Technical inspection mark (1922-1939)	Poland
shield	Technical inspection mark (1922-1939)	Poland

MAUSER MILITARY RIFLE MARKINGS

Symbol	Meaning	Country
	Arsenal rework mark (Z = *Zbrowania*, i.e., Arsenal; this one was No.4, at Krakow)	Poland
	Kingdom of Prussia national emblem	Prussia
	Indicates captured foreign weapons	Russia/USSR
	Cyrillic S, abbreviation for Србија (Serbia)	Serbia
	A *chakra*, a Thai national symbol	Siam (Thailand)
	A variant of the *chakra*, having a royal crown in the center	Siam (Thailand)

MAUSER MILITARY RIFLE MARKINGS

Symbol	Meaning	Country
	A Thai government property mark	Siam (Thailand)
	National symbol found on arms of the Nazi puppet state, 1939-45	Slovakia
	Military[?] property mark, first used in the 1930s	Spain
	Rifle was inspected, zeroed and test-fired	Sweden
B	Boden depot mark	Sweden
C	Carl Gustaf depot mark	Sweden
CB	Repair mark, Carlsborg Arsenal	Sweden
G	Gotland depot mark	Sweden
Gg	Göteborg depot mark	Sweden
K	Karlsborg (old spelling of Carlsborg) depot mark	Sweden

MAUSER MILITARY RIFLE MARKINGS

Symbol	Meaning	Country
♛ O	Ostersund depot mark	Sweden
♛ S	Stockholm depot mark	Sweden
(toughra)	*Hanı Abdülaziz bin Mahmud elmuzaffirü dâima* (Khan Abdülaziz, son of Mahmud; the Ever Victorious) Toughra of Sultan Abdülaziz	Turkey (Ottoman Empire)
(toughra)	*Hanı Mehmet Murad bin Abdülmecid elmuzaffirü dâima* (Khan Mehmet Murad, son of Abdülmajid; the Ever Victorious) Toughra of Sultan Murad V	Turkey (Ottoman Empire)

MAUSER MILITARY RIFLE MARKINGS

Symbol	Meaning	Country
	Hanı Mehmet Murad bin Abdülmecid elmuzaffirü dâima (Khan Mehmet Murad, son of Abdülmajid; the Ever Victorious) Toughra of Sultan Murad V	Turkey (Ottoman Empire)
	Hanı Abdülhamid bin Abdülmecid elmuzaffirü dâima, Elgazi (Khan Abdülhamid, son of Abdülmajid; the Ever Victorious, The Warrior Champion) Toughra of Sultan Abdülhamid II	Turkey (Ottoman Empire)

Symbol	Meaning	Country
	Hanı Mehmet bin Abdülmecid elmuzaffirü dâima, Reşad (Khan Mehmet, son of Abdülmajid; the Ever Victorious, He of Integrity) Toughra of Sultan Mehmet V	Turkey (Ottoman Empire)
	Hanı Mehmet Vahidüddin bin Abdülmecid elmuzaffirü dâima, Elgazi (Khan Mehmet Vahidüddin, son of Abdülmajid; the Ever Victorious, The Warrior Champion) Toughra of Sultan Mehmet VI	Turkey (Ottoman Empire)

MAUSER MILITARY RIFLE MARKINGS

Symbol	Meaning	Country
☾	Crescent moon	Turkey (Ottoman Empire and Republic)
☪	Crescent moon and star	Turkey (Ottoman Empire and Republic)
ATF	The monogram ATF, *Ankara Tüfek Fabrikası* (Ankara Rifle Factory) Arsenal mark (c.1949-c.1954)	Turkey
ROU	The monogram ROU (*República Oriental del Uruguay*)	Uruguay
BTZ	Serbian Cyrillic letters VTZ (*VojnoTechnichiki Zavod* = Military Technical Factory) logo, Kragujevac arsenal	Yugoslavia

MAUSER MILITARY RIFLE MARKINGS

Symbol	Meaning	Country
𝒯.𝓡.𝒵.5	T.R.Z. 5 *Tehnički Remont Zavod* (Technical Repair Factory) No. 5	Yugoslavia
(Ц А)	Stock inspection or acceptance mark	Yugoslavia

NOTES ON THE EMBLEMS AND THEIR MEANINGS

Bosnia-Herzegovina ---

Though most commonly associated with France, the *fleur-de-lys* (also *fleur-de-lis*) appears as a symbol here and there around Europe (and in Québec and Louisiana); in this instance it is part of the crest of Bosnia-Herzegovina. It is a stylized lily of the valley, consisting of three iris leaves encircled by a ribbon. The *fleur-de-lys* is found on various Yugoslav incarnations of M98 Mausers acquired by Bosnia-Herzegovina, one of the former republics of the Yugoslav federation, after it gained independence when Yugoslavia disintegrated in 1992.

China ---

The *Kuomintang* (also spelled *Guomindang*) is the Chinese Nationalist Party. This was the party of Sun Yat-sen and Chiang Kai-shek. The Nationalists were defeated by Mao Tse-tung's Communists in 1949 and fled to the island of Taiwan, where they re-established the Republic of China. The sunburst design was adopted by the *Kuomintang* in 1928 and is still used by the R.O.C.

The *swastika* (from Sanskrit *svasti*, meaning, approximately, "well-being" or "beneficial"), although now for obvious reasons identified principally with the Nazis, is a very ancient symbol. It is found all over the world and in cultures as different and geographically remote from each other as Egyptian Coptic Christianity, pre-Christian Germanic and Slavic religions, and American Indian societies. One theory is that the hooked cross represents the wheel of the sun, but this is not subject to proof as the symbol is far too old and widespread to be pigeonholed quite so easily.

The swastika is a common symbol in Buddhism, imported from India into China and elsewhere; it is sometimes found on his chest in paintings and statues of the Buddha, and represents his heart. In Buddhist tradition, as shown in the illustration, the swastika typically faces left and is foursquare, unlike the Nazi version, which faces right and is tilted 45° to the left. For our purposes the notable thing is that the swastika was the factory logo for Arsenal No. 21, which was part of the arms production facilities at Hanyang.

For information on the Zhejiang Iron Works and the Chinese Model 77 rifle please see the chapter on **Inscriptions in Non-Western Languages**.

Czechoslovakia ---

Arms made by Czechoslovak factories for and actually accepted by the Czech armed forces were stamped with an acceptance mark by the *Vojenský technický ústav* (Military Technical Institute, abbreviated VTU; in 1936 it

MAUSER MILITARY RIFLE MARKINGS

was renamed *Vojenský technický a letecký ústav*, Military Technical and Aviation Institute, abbreviated VTLU.) This acceptance mark initially took the form of a monogram of the letters ČSR, for *Republica Československa* (Czechoslovak Republic), used in 1918-22. In 1922 this was changed to a Bohemian lion in a rectangle, which changed again in 1923 to the letter **E** followed by the lion in a circle, followed by the last two digits of the year of acceptance. In the years 1924-36 the circle around the lion was omitted. In 1937 a number from 1 to 8 was added before the **E**, indicating the location of the VTLU that accepted the weapon for service. The VTLU code numbers and locations were as follows; place names in parentheses are those more familiar to English-speakers:

E 1 = Plzeň (Pilsen)
E 2 = Adamov
E 3 = Brno
E 4 = Povážská Bystrica
E 5 = Vlašim
E 6 = Semtín
E 7 = Strakonice
E 8 = Praha (Prague)

At its height the pre-World War II Czechoslovak Army had twelve divisions. Some Czech weapons have the division HQ code stamped before the two-numeral date of acceptance; this can be recognized by the prefix **S**. For example, S 923 indicates property of the division headquartered in Bratislava, accepted in 1923.

The division HQ codes are listed below. Once more, the place names in parentheses are those more familiar to English-speakers, excepting Užhorod, whose alternate spelling *Užgorod* is given because it sometimes appears on Czechoslovakian items.

MAUSER MILITARY RIFLE MARKINGS

S 1 = Praha (Prague)
S 2 = Plzeň (Pilsen)
S 3 = Litoměřice
S 4 = Hradec Králové
S 5 = České Budějovice (Budweis)
S 6 = Brno
S 7 = Olomouc
S 8 = Opava
S 9 = Bratislava
S 10 = Bánská Bystrica
S 11 = Košice
S 12 = Užhorod (Užgorod)

Dutch East Indies ---

The Netherlands government, for a number of reasons, preferred to rule the Dutch East Indies indirectly to a great extent. The *Indische Ondernemersbond*, "Indies Entrepreneurs' Organization" (or to give it the more descriptive translation the British used, "Federation of Indian Industry and Commerce") was a hybrid organization, neither entirely private nor entirely governmental, consisting of representatives of seventeen different commercial and agricultural sectors (rubber, tobacco, sugar, etc.) that acted as an advisory and information board for the Royal Dutch government. The IOB members had considerable control over the internal affairs of their possessions, which were almost fiefdoms held from the Netherlands government. Given the circumstances, and its members' broad local authority and powers, it is not especially surprising that there should be IOB-marked Mausers.

MAUSER MILITARY RIFLE MARKINGS

The Golden Age of the IOB was the 1920s and '30s; the Japanese invasion in 1942 and occupation during World War II brought economic and political ruin to the IOB as well as to the Netherlands government in the Dutch East Indies. In 1949 international pressure, native rebellion, and post-war exhaustion compelled the Netherlands to grant independence to the Dutch East Indies, which then became the Republic of Indonesia. The Dutch managed to hold West Irian (the western part of the island of New Guinea), a tiny piece of their enormous former colony, until 1962, when the U.N. forced them to surrender the area to Indonesia, which had invaded and occupied it. The region is Indonesian territory to this day.

Ethiopia ---

"Conquering Lion of the Tribe of Judah" was a title of the Ethiopian emperor; it refers to descent claimed from an amorous dalliance between the biblical King Solomon and Belkis, Queen of Sheba. The title is most commonly associated with the last Ethiopian emperor, Haile Selassie I.
This emblem can be found facing either left or right, but on Ethiopian Mausers it generally faces left.

Germany ---

German military arms inspectors before the end of the First World War in November 1918 had individual acceptance stamps assigned to them consisting, generally, of the first letter of their last name in *Fraktur* type (see **Alphabets**), surmounted by a crown. Many different letters can be found, and there are several different styles of crowns. As there were many inspectors with different surnames beginning with the same first letter, there were a number of people using a particular letter at any given time; therefore, a number of variations exist. This did not cause confusion at the time, however, because exact copies of all inspectors' markings were on file with the military inspectorate.

During World War I the German government faced shortages of most materials, eventually resulting in the use of non-standard parts in some rifles. These were either parts that had been rejected for some reason, or perfectly acceptable parts from non-government manufacturers. Such rifles were marked with a star on the receiver, hence the term *Sterngewehr* ("star rifle") sometimes used to describe them. There are at least two varieties of the star marking.

The *Inspektion für Waffen- und Gerät* (I.W.G.) was the German military office responsible for inspection and acceptance of small arms and equipment between the two World Wars, meaning, to all intents and purposes, from 1920 to 1935. It was the predecessor of the *Waffenamt*. [For a discussion of the *Waffenamt* see the chapter on **German Wartime Codes**.]

Use of the swastika (German *Hakenkreuz*, literally "hooked cross") as a Nazi symbol is so familiar that no discussion of it is necessary. The significance of the SS (for *Schutz-Staffeln*, "Protection Squad") symbol, the *Sig-Runen*,[*] is less well known. Often ---though wrongly--- referred to as "lightning bolts" because of their appearance, they are actually derived from the Runic alphabet used by some of the northern Teutonic tribes in ancient times, and were adopted by the Nazis as part of their New Age-like neo-pagan revival. The *Sig-Runen* are merely two Runic

---

[*] This is not a misprint for "*Sieg*": the term really is *Sig-Runen*. Though *Sieg-Runen* sometimes appears even in post-war German publications, it is not, strictly speaking, correct.

esses, for *SS*. A single S-rune was used as a symbol by the *Deutsches Jungvolk*, the organization for children too young for the *Hitler-Jugend* (Hitler Youth), and occasionally by the *Hitler-Jugend* as well, but that has no connection with its use on Mauser rifles as described in the chart.

The Death's Head (*Totenkopf*), like the swastika and runes, is an old Germanic symbol the Nazis adopted. Popular with soldiers' groups during the late Middle Ages, it originally signified a unit's loyalty to its employer unto death. In the age of mercenary armies this was an effective advertising gimmick

Almost every imaginable (and some unimaginable) Third Reich item has been faked for the collector market, and weapons markings are no exception. Anything purported to have genuine Third Reich markings should be examined with the utmost care and double- and triple-checked against all other possible indicators in order to assure its genuineness, and in general should be regarded as a fake unless proven otherwise.

Iraq ---

I have sometimes seen the marking ⟨℡⟩ referred to as indicating a rifle used by Iraq's Republican Guard. Absent any documentation for it I have doubts about this attribution for several reasons, chief among them being the following:

First, the Republican Guard (though now to all intents and purposes an elite army unit) is actually the paramilitary arm of the Iraqi branch of the *Ba'ath* (meaning "renaissance") Party, the Arab socialist organization that, in its various permutations, has run Iraq and Syria for decades. The Republican Guard was formed at the end of 1963, after a series of coups d'état, as a palace guard and goon squad for

MAUSER MILITARY RIFLE MARKINGS

the Party leadership. Personal weapons are important status symbols in Arab culture, and for such a politically important and pampered group to have been armed, even for ceremonial functions, with what were already aging, obsolete bolt action rifles of numerous models and from many countries makes little sense; these are the sorts of weapons issued to cannon-fodder conscripts fresh from the villages, not to an elite unit.

Second, in Arabic "Republican Guard" is *Haras Jumhūrī* (حرس جمهوري), so I suggest that if there is an "alphabetical" marking it would be ح ج though, certainly, a simple ج for *Jumhūrī* (جمهوري), "Republican", is possible. ج is not likely to have been used as it could also stand for *Haras Qawmī* (حرس قومي), Iraq's National Guard. If Republican Guard weapons are specifically marked at all one would expect them to have a more elaborate device than a small single letter in a triangle.

Consequently, I believe the ج is more likely an abbreviation for the word *jaysh* (جيش), "army", in الجيش الشعبي, *al-Jaysh ash-Sha'abī*, literally "the People's Army". Sometimes confusingly referred to in English-language publications as the "Popular Militia" or "People's Militia", *al-Jaysh ash-Sha'abī* is actually the Ba'ath Party militia.

I may be wrong in this conclusion, i.e., attributing the marking to "army", but it is my opinion as of the time of this writing.

Lithuania ---

The chart shows the "Columns of the Gediminas family", a famous Lithuanian dynasty. This emblem appears on the Gediminas family's coat of arms after 1397; their subsequent dynastic offshoots, the Kęstutis and Jogalla families, effectively adopted it as the Lithuanian

MAUSER MILITARY RIFLE MARKINGS

national symbol. It was affirmed as such after Lithuania achieved independence from Russia in 1918, and was used as the insignia of the army, navy, air force, police, and other national organizations. After their annexation of Lithuania in 1940 the Soviets banned the use and display of the Columns. In 1988 the symbol was revived by the *Sajūdis* political movement, which was instrumental in Lithuania's rebirth as an nation in 1991. Since independence the "Columns of the Gediminas family" has once more become the symbol of the Lithuanian military, police, National Olympic Committee, and other state institutions.

Manchuria ---

From 1932 to 1945 this bleak, industrial, coal-mining region in northeastern China was a Japanese puppet state called *Manchukuo* ("Manchuland"), nominally under the rule of the hapless last emperor of China, Henry Pu-yi (b.1906, d. 1967). Manchuria was the site of the Mukden arsenal, maker of Mauser rifles and other useful appliances. Mukden is now called *Shenyang* and is the capital of *Liaoning* Province, which was formerly called *Kwantung*.

Persia ---

On the butt stock of Persian Mausers there is generally a serial number in Persian numerals accompanied by a Persian letter. Attempting to match the Persian letter with a Latin one is futile, as in many instances there is no equivalent letter; in any event there are more Persian letters than Latin ones, and an exact correspondence is therefore

MAUSER MILITARY RIFLE MARKINGS

not possible. For a discussion of these matters see the chapter on **Alphabets**.

The word *piyādeh* ("Infantry") is also found on many of the M98/29 Mausers made for Iran; this is shown in the chart.

Several different inscriptions are to be found on the side rails of the various Persian Mauser models, for translations of which see the appropriate items on the foregoing chart.

Prussia ---

Note that the Prussian eagle is not the same as the imperial eagle of the German Empire. (See the chapter on **Coats of Arms**). When they were unified as the Second Reich in 1871 ---the First Reich was the empire of Charlemagne, reckoned from his coronation as Emperor of the Romans [*sic*] on Christmas Day in 800 A.D.--- the various German states kept their rulers and local emblems, which is why the monograms of the kings of places such as Bavaria and Saxony appear on military equipment as late as World War I. The rulers of the individual states were, however, subordinate to the king of Prussia in his capacity as *Kaiser* (emperor) of the Reich. *Kaiser*, by the way, is a corruption of "Caesar", as is the Russian title *tsar*.

Mausers bearing the Prussian eagle marking can date from earlier than September 1871, or can be attributed to military units of the Prussian kingdom rather than the *Reich*. The Prussian kingdom ended with the Second Reich when Germany lost World War I in November 1918 and the king/Kaiser fled to the Netherlands.

MAUSER MILITARY RIFLE MARKINGS

Russia ---

Because of its small size and typically careless stamping the Russian capture/reissue mark for foreign weapons (found on the side of the receiver ring and elsewhere) usually resembles a deformed **X**. It is in fact a pair of crossed Mosin-Nagant rifles.

Siam ---

In Thai mythology the *chakra* (a Sanskrit word meaning, literally, "wheel") was a weapon of the gods, and is found as a marking in several variations on Siamese Mausers. This item can be seen in various forms throughout the Orient even today as a traditional martial arts weapon.

Slovakia ---

As found on Mausers this refers to the Nazi puppet state created by the Germans after they annexed the remnants of Czechoslovakia in 1939, and not to the present nation which arose from the dissolution of Czechoslovakia in 1993. The wartime version of Slovakia lasted only until the end of the Second World War in Europe in May 1945; its head of state was a Roman Catholic priest, Josef Tiso,[*] who was hanged by the Czechs in 1947 after being convicted of committing war crimes. The mounds on the crest represent the three mountain ranges of Slovakia: Fatra, Matra and Tatra, surmounted by a patriarchal cross.

---

[*] Not to be confused with Josip Broz Tito, the Communist partisan commander and post-World War II leader of Yugoslavia.

MAUSER MILITARY RIFLE MARKINGS

Turkey ---

Turkish Mausers made or reworked during the time of the Ottoman Empire and the early years of the Republic originally have markings in the Arabic-derived alphabet in which Turkish was written at the time. For a thorough discussion of this matter see the **Languages** section.

Arabic numbers were also the norm for Turkish during this period. Often, though incorrectly, these are called "Farsi" numbers in some publications: Farsi, i.e., Persian, numbers differ considerably from the Arabic for the numerals 4, 5 and 6. (See the **Numbers** chapter elsewhere in this book). Although the writing is read from right to left, the numbers are read left to right. This phenomenon is also discussed in the **Numbers** chapter.

Dates on Turkish rifles prior to adoption of the Western calendar in December 1925 are in the Islamic system. For a detailed discussion see the chapter on **Calendars and Dating**.

The calligraphic design commonly found on Turkish Mausers of the Ottoman period is the *toughra* (in Turkish, *tuğra*). This is an ornate design individual to each sultan and consists of the title *hanı* (the Turkish form of *khan*), the sultan's name, the Arabic word *bin* ("son [of]"), the sultan's father's name, and the title *elmuzaffirü dâima* (the Turkish form of the Arabic *al muzaffir da'ima*, meaning "the ever victorious"). This mix of languages shows the eclectic nature of the Ottomans and their empire: *khan* was originally a Mongol title meaning, approximately, "ruler" and the last title is, as noted, Arabic.

The toughra is supposed to have originated with Sultan Murad I who, when signing the Treaty of Ragusa (now Dubrovnik, Croatia) in 1365, dipped his fingers in an inkwell and scrawled a signature. This design developed

MAUSER MILITARY RIFLE MARKINGS

into a standardized form having three vertical lines and a swirl of calligraphy ending in a flourish to the left. As they are written in Arabic script toughras are read from lower right to upper left. Although toughras of many of the sultans look alike at first glance, patience and a good magnifying glass or jeweler's loop can sometimes be used to untangle them.

A smaller calligraphic device is sometimes found above and to the right of the toughra. This consists of an additional title or attribute (the Turkish term is *ünvan*, from the Arabic word *'unwān*, meaning "address" or "title"), which also is unique to each sultan, though not every sultan had one. The *ünvan* can be quite useful for sorting out the six Mehmets, five Murads, four Mustafas, etc. Like the toughra this device, because of its intricacy, is sometimes difficult to read even by people with a good knowledge of Arabic script. For those interested in knowing them the titles for those sultans of the "Mauser Era" are set forth in the following chart. Though all these titles are of Arabic origin they are given here in their Turkish forms:

Sultan	Reigned	Title	Meaning
Abdülaziz	1861-1876 A.D. (1277-1293 A.H.)	---	---
Murad V	1876 A.D. (1293 A.H.)	---	---
Abdülhamid II	1876-1909 A.D. (1293-1327 A.H.)	Elgazi	Warrior for the Faith (i.e., Islam)
Mehmet V	1909-1918 A.D. (1327-1336 A.H.)	Reşad	Man of integrity

MAUSER MILITARY RIFLE MARKINGS

Sultan	Reigned	Title	Meaning
Mehmet VI	1918-1922 A.D. (1336-1341 A.H.)	Vahidüdin	Unequalled in the Faith

At the beginning of World War I Mehmet V also adopted the title *Elgazi*, for obvious reasons. Possibly because his toughra already had the *ünvan* "Reşad", Mehmet V did not add *Elgazi* to it. When he succeeded his brother in 1918, while the war was still going on, Mehmet VI also assumed the title *Elgazi*.

The word *Elgazi* (the Turkish form of the Arabic words *al ghāzi*) is interesting in that it originally meant "raider" or "invader", but gradually came to mean a warlord or warrior champion, with the implied meaning of "warrior for Islam". The German language still has the slang word *Razzia*, meaning a police raid, from exactly the same origin.

"Vahidüddin" was actually part of Mehmet VI's name, and thus is technically not an *ünvan*.

I have furnished drawings of five of these toughras in this chapter. Abdülaziz's toughra might not be found on rifles, but is included in the event one does turn up that bears it. Murad V reigned so briefly (May 30 – August 31, 1876) that his toughra did not appear on much of anything, including, probably, Mauser rifles, but is also included for the sake of completeness.

The crescent moon, with or without a star, is an old Islamic symbol found all over the Muslim world to this day. The Turks adopted it in the Middle Ages, and it became so identified with them that they retained it as their national symbol even after abolishing the empire and sultanate in the 1920s. The star did not become a regular addition to the Turkish crescent until the late 1790s, though it had appeared from time to time before then with differing numbers of points: eight, or six, or five. The crescent typically faces right on most things Turkish, but on their Mausers it almost always faces upward.

MAUSER MILITARY RIFLE MARKINGS

GERMAN WORLD WAR II CODES

During World War II the standard rifle of the German armed forces, the *Wehrmacht*, was the Model Kar98k rifle. The weapon was produced in the millions --- 12,000,000, by some accounts--- by numerous arsenals, and often contains components made by subcontractors of small parts, military equipment, etc.

Armaments production is a military secret in most cases, and in order to conceal various aspects of their arms industries the German government assigned codes to producers of weapons and their parts. There were a great many codes, and some were changed from time to time; a complete record of them has yet to be compiled, but with the continuing opening of sources in the former East Bloc knowledge is flowing more freely than at any time since before the war. Some day all the codes may well be cracked. The following are those known at the time of this writing. Firms which made parts, rather than entire rifles as such, are indicated here by an *; the asterisk was **not** part of the code. For additional maker codes see the **Waffenamt** section later in this chapter.

By way of explanation, "Brünn" (also spelled "Bruenn") is the German name for the Czech city Brno; "Bystrica" is the German name for Bystřice, also a town in what was Czechoslovakia and is now the Czech Republic.

Code	Maker and Location
a	Nahmatag, Nähmaschinenteile, AG; Dresden
ar	Mauser-Werke, AG; Werk Borsigwalde, Berlin-Borsigwalde
avk *	Ruhrstahl, AG, Presswerk Brackwede bei Bielefeld
awz *	Wille, Eduard, Werkzeugfabrik, Hammerwerke; Wuppertal-Cronenburg
ax	Feinmechanische Werke GmbH; Erfurt (a/k/a ERMA, this firm used the code 27 until 1940)

MAUSER MILITARY RIFLE MARKINGS

Code	Maker and Location
ayf *	B. Geipel GmbH Waffenfabirk (ERMA); Erfurt
bcd	Gustloff Werke, Werk Weimar; Weimar
bcd/ar	Indicates production by the above two makers
bnz	Steyer-Daimler-Puch AG; Steyr, Austria
bpr *	Johannes Grossfuss Metall-und Lakierwarenfabrik; Döbeln
brg *	H.W. Schmidt Metallwarenfabrik; Döbeln
BSW*	Berlin-Suhler-Waffen
btd *	Richard Sieper & Söhne
byf	Mauser-Werke, AG; Oberndorf am Neckar
bys	Ruhrstahl, AG; Gusstahlwerke; Witten/Ruhr
ce	J.P. Sauer und Sohn, Suhl
ch *	Fabrique National; Herstal, Belgium
crv *	Fritz Werner AG Maschinen-und Werkzeugfabrik; Werk II, Berlin
cyw *	Sächische Gusstahlwerke Döbeln AG, Stahlwerke Freital/Sachsen
dlv *	Deutsche Edelstahlwerke, Werk Remscheid
dot	Waffenwerke Brünn AG, Brno, Czechoslovakia
dou	Waffenwerke Brünn AG, Bystrica, Czechoslovakia
duv	Berlin-Luebecker Maschinenfabriken, Werk Luebeck (its code was 237 until 1940)
dwc *	Bome & Co Dr Ing Werksleitung Lüdenscheid, Werke Minden Westliche
e *	Hermann Kohler, AG Maschinenfabrik; Altenberg, Thüringen
eeu *	Lieferungsgemeinschaft Westthüringen Werkzeug- und Metallwarenfabriken GmbH, Schmalkalden
fxo *	C G Haenel Waffen-und Fahrradfabrik; Suhl
gba *	Adolf von Brauke AG, Gusstahl-Draht-und Seilwerke; Ihmertenback
Geco	Gustav Genschow & Co AG; Berlin
ghn *	Carl Ullrich & Co Metallwarenfabrik; Oberschönau/Thüringen
gqm *	Lock & Hartenberger Metallwarenfabrik; Idar-Oberstein
guo *	NV Nederlandische Maschinenfabrik, Artillerie Inrichtlingen; Zaandam
i *	Elite Diamantwerke AG, Siegmar-Schonau; Chemnitz
jvh *	Metallwaren-Waffen- und Maschinenfabrik A OG, Budapest (a/k/a FEG)

MAUSER MILITARY RIFLE MARKINGS

Code	Maker and Location
jwh *	Staatliche Waffenfabrik; Châtellerault, France
K *	Luck & Wagner; Suhl/ Thüringen
LU *	?
lxr *	Dianawerk Mayer & Grammelspacher; Rastatt/Baden
ofh *	Metall- und Kunzharzwerk GmbH, Komotau
ouj *	VDM Luftfahrtwerke A-G, Werk Friedland, Bezirk Breslau
q *	Julius Kohler; Limbach/Sachsen
qnw *	?
s *	Dynamit A-G (formerly Alfred Nobel & Co.), St.-Lambrecht
S/27	Erfurter Maschinenfabrik (ERMA)
S/27G	Erfurter Maschinenfabrik (ERMA)
S/42	Mauser-Werke, AG, Oberndorf am Neckar (early World War II)
S/42G	Mauser-Werke, AG, Oberndorf am Neckar (1935 only?)
S/42K	Mauser-Werke, AG, Oberndorf am Neckar (1934 only?)
S243	Mauser-Werke, AG; Werk Borsigwalde, Berlin-Borsigwalde
S243G	Mauser-Werke, AG; Werk Borsigwalde, Berlin-Borsigwalde
S/147	J.P. Sauer und Sohn, Suhl
S/147G	J.P. Sauer und Sohn, Suhl
S/147K	J.P. Sauer und Sohn, Suhl
S/237	Berlin-Luebecker Maschinenfabriken, Werk Luebeck
svw MB *	Mauser-Werke, AG, Oberndorf am Neckar (late World War II)
swp *	Waffenwerke Brünn AG, Brno, Czechoslovakia
27	Erfurter Maschinenfabrik (ERMA) (until 1940, when the code changed to ax)
42	Mauser-Werke, AG, Oberndorf am Neckar (an early code)
147	J.P. Sauer und Sohn, Suhl
237	Berlin-Luebecker Maschinenfabriken, Werk Luebeck (its code was changed in 1940 to duv)
243	Mauser-Werke, AG; Werk Borsigwalde, Berlin-Borsigwalde

MAUSER MILITARY RIFLE MARKINGS

Code	Maker and Location
337	Gustloff Werke, Werk Weimar, Weimar
660	Steyer-Daimler-Puch AG, Steyr, Austria
945	Waffenwerke Brünn AG, Brno, Czechoslovakia

WAFFENAMT MARKINGS

In Germany during the 1930s and 1940s, until the end of World War II, quality control of weapons, weapons parts, munitions, and other related military materials was the responsibility of the *Waffenamt* (literally "Weapons Administration), the successor to the Reichswehr's small arms and equipment inspectorate of the inter-war years, the *Inspektion für Waffen- und Gerät.*

Each inspector was assigned an individual number, which he and his assistants stamped on items accepted for military use. Inspectors did not generally stay at the same facility throughout their service, so the same inspector's mark can be found on products from several factories. The *Waffenamt* mark consisted of an eagle and a number; some inspection stamps included the letters **WaA** (for *Waffenamt*) and some do not. I have not included the letters **WaA** with the numbers below because they do not make a difference in the inspector's designation. [For illustrations of representative *Waffenamt* marks see the **Emblems** chapter.] The Germans had a slang term for the eagle on these stamps: *Plattgeier*, meaning "flat vulture". Some translate this as "bankrupt vulture" as *platt* is indeed a slang word for "bankrupt"; to me this makes less sense than "flat" in this context.

MAUSER MILITARY RIFLE MARKINGS

The second set of numbers (in some cases letters, or letter and numbers) shown in **boldface** in the following chart is the factory code for the facility where the inspector served at the time.

The list below consists of *Waffenamt* inspectors' numbers known ---or believed--- to have been involved in the inspection of Mauser rifles, parts or accessories such as magazines and grenade launchers. Markings pertaining to bayonets, slings, ammunition pouches, etc., are not included in the following list as they are outside the scope of this book.

No complete record detailing all *Waffenamt* inspectors' marks is known to have survived the war, and it is possible, indeed probable, that more attributable markings will surface as the years pass and information continues to emerge from behind the former Iron Curtain. Let us hope so.

WaA mark and Code	Factory and Location	Period
1	Berlin-Suhler Waffenwerke (Suhl)	1937-39
1 **337**	Berlin-Suhler Waffenwerke (Suhl)	1939-40
1 **bcd**	Gustloff-Werke (Weimar)	1941-45
4	Berlin-Suhler Waffenwerke (Suhl)	1937-39
4	Berlin-Suhler Waffenwerke (Suhl)	1937-39
18 **bcd**	Gustloff-Werke (Weimar)	?

MAUSER MILITARY RIFLE MARKINGS

WaA mark and Code	Factory and Location	Period
26 S/237	Berlin-Lübecker Maschinenfabriken (Lübeck)	1936-38
26 237	Berlin-Lübecker Maschinenfabriken (Lübeck)	1938
26 S/243	Mauser-Werke (Borsigwalde)	1938
26 243	Mauser-Werke (Borsigwalde)	1938-40
26 ar	Mauser-Werke (Borsigwalde)	1941-44
37 147	Sauer & Sohn (Suhl)	1939-40
37 ce	Sauer & Sohn (Suhl)	1941-44
49 S/147K	Sauer & Sohn (Suhl)	1934
49 S237	Berlin-Lübecker Maschinenfabriken (Lübeck)	1936
49	Mauser-Werke (Borsigwalde)	?
51	Erma (Erfurt)	?
63 S/42	Mauserwerk (Oberndorf)	1935-38
63 42	Mauserwerk (Oberndorf)	1938-39
63 945	Waffenwerk Brünn (Brno, CZ)	1940

MAUSER MILITARY RIFLE MARKINGS

WaA mark and Code	Factory and Location	Period
63 **dot**	Waffenwerk Brünn (Brno, CZ)	1941-45
63 **swp**	Waffenwerk Brünn (Brno, CZ)	1945
77 **bnz**	Steyr-Daimler-Puch (Steyr, Austria)	1940-44
80 **dou**	Waffenwerk Brünn (Bystrica, CZ)	1941-45
108 **S/42**	Mauserwerk (Oberndorf)	1935
114 **S/237**	Berlin-Lübecker Maschinenfabriken (Lübeck)	1936
114 **S/147**	Sauer & Sohn (Suhl)	1934-35
115 **S/147**	Sauer & Sohn (Suhl)	1934
116 **S/147**	Sauer & Sohn (Suhl)	1934-36
132	Erma (Erfurt)	1943
134 **bcd**	Gustloff-Werke (Weimar)	1943
135 **S/42**	Mauserwerk (Oberndorf)	1935
135 **byf**	Mauserwerk (Oberndorf)	1941-45
135 **qnw**	?	?

MAUSER MILITARY RIFLE MARKINGS

WaA mark and Code	Factory and Location	Period
140 **ch**	Fabrique National (Liège, Belgium)	?
211 **S/42**	Mauserwerke (Oberndorf)	1935
211	Mauser-Werke (Borsigwalde)	?
214 **S/147**	Sauer & Sohn (Suhl)	1935-38
214 **147**	Sauer & Sohn (Suhl)	1938
214 **237**	Berlin-Lübecker Maschinenfabriken (Lübeck)	1939-40
214 **237**	Berlin-Lübecker Maschinenfabriken (Lübeck)	1939-40
214 **duv**	Berlin-Lübecker Maschinenfabriken (Lübeck)	1940-44
214 **qve**	Berlin-Lübecker Maschinenfabriken (Lübeck)	1945
214	Erma (Erfurt)	1936
217 **S/243**	Mauser-Werke (Borsigwalde)	1935-37
218 **bcd**	Gustloff-Werke (Weimar)	?
221 **e**	Herman Köhler (Altenburg)	1942

MAUSER MILITARY RIFLE MARKINGS

WaA mark and Code	Factory and Location	Period
241 **S/42**	Mauserwerke (Oberndorf)	1935
280 **S/27**	Erma & Geipel (Erfurt)	1935-37
280 **27**	Erma (Erfurt)	1938-40
280 **ax**	Erma (Erfurt)	1940-41
280 **S/243**	Mauser-Werke (Borsigwalde)	1938
280 **ce**	Sauer & Sohn (Suhl)	1942-44
359 **S/147**	Sauer & Sohn (Suhl)	1935-38
359 **147**	Sauer & Sohn (Suhl)	1938-40
359 **ce**	Sauer & Sohn (Suhl)	1941-44
359	Erma (Erfurt)	?
497 **bpr**	J. Grossfuss (Döbel)	?
497 **brg**	H. W. Schmidt (Döbel)	?
607	Waffenwerke Brünn (Bystrica, CZ)	1940
607 **dou**	Waffenwerke Brünn (Bystrica, CZ)	1941

MAUSER MILITARY RIFLE MARKINGS

WaA mark and Code	Factory and Location	Period
623 **660**	Steyr-Daimler-Puch (Steyr, Austria)	1939-40
623 **bnz**	Steyr-Daimler-Puch (Steyr, Austria)	1940-45
655 **42**	Mauserwerke (Oberndorf)	1938-41
655	Sauer & Sohn (Suhl)	1940
742 **S/27**	Erma & Geipel (Erfurt)	1935
749 **337**	Berlin-Suhler Waffenwerke (Suhl)	1940
749 **bcd**	Gustloff-Werke (Weimar)	1941-44
A68 **adb**	Markneukirchner Metallbearbeitungs, GmbH (Markneukirchner)	?
A80 **dou**	Waffenwerke Brünn (Bystrica, CZ)	1942, 1944-45
B43 **aye**	Olympia Büromaschinenwerke (Erfurt)	1943-45
B44	?	?
B92	Grohman & Sohn (Würbenthal Süd)	1944-45

MAUSER MILITARY RIFLE MARKINGS

WORLD WAR II GERMAN SUBCONTRACTOR CODES

In addition to the principal Mauser rifle factories where entire weapons were produced or assembled (Mauser, J. P. Sauer & Sohn, etc.) there were many subcontractors who provided parts to the principals. These companies had codes to disguise their identities, just as did the major producers. Many of these subcontractors' codes are known now, though many remain obscure. The reader will note that in some instances a primary manufacturer will also be listed here as a subcontractor, and I have done this because some parts (the floorplate, for example) can turn up on a rifle made by a different primary manufacturer, and yet the rifle will still be entirely authentic and original.

There are, obviously, a number of letters which appear on many different types of Mausers and which mean many different things. For this reason the reader must use a certain amount of deduction in order to attribute, for example, a **K**, to the proper meaning (subcontractor code, *Kürassier* regiment, and so on); in practice this is not especially difficult.

Code	Subcontractor	Location
a	Nahmatag, Nähmaschinenteile AG	Dresden, Germany
ar	Mauser Werke	Borsigwalde, Germany
avk	Ruhrstahl, A.G.	Brackwede bei Bielefeld, Germany
ax	Feinmaschinische Werke, GmbH	Erfurt, Germany

MAUSER MILITARY RIFLE MARKINGS

Code	Subcontractor	Location
ayf	ERMA	Erfurt, Germany
bpr	Johannes Grossfuss Metall- und Lakierwarenfabrik	Döbeln, Germany
brg	H.W. Schmidt Metallwarenfabrik	Döbeln, Germany
btd	Richard Sieper & Söhne	?
ce	J. P. Sauer & Sohn Gewehrfabrik	Suhl, Germany
cyw	Sächsische Gussstahlwerke Dohlen, AG	Freital/Sachsen, Germany
dhs	Schöniger Maschinenfabrik, GmbH	Schönigen, Germany
dlv	Deutsches Edelstahlwerke AG, Werke Remscheid	Remscheid, Germany
dwc	Böme & Co. Dr. Ing. Werksleitung Lüdenscheid	Minden (Westfalen), Germany
e	Hermann Kohler AG Maschinenfabrik	Altenburg (Thüringen), Germany
eeu	Lieferungsgemeinschaft Westthüringen Werkzeug- und Metallwarenfabriken GmbH	Schmalkalden

MAUSER MILITARY RIFLE MARKINGS

Code	Subcontractor	Location
fxo	C.G. Haenel Waffen- und Fahrradfanrik	Suhl, Germany
gba	Adolf von Braucke AG, Gussstahl-Draht- und Seilwerke	Ihmerterback bei Westig (Westfalen), Germany
ghn	Carl Ullrich &Co. Metallwarenfabrik	Oberschonau (Thüringen), Germany
gqm	Lock & Hartenberger Metallwarenfabrik	Idar-Oberstein, Germany
guo	NV Nederlandische Maschinenfabrik Artillerie Inrichingen	Zaandam, Netherlands
i	Elite Diamantwerke AG	Siegmar-Schonau bei Chemnitz, Germany
jvh	Metallwaren- Waffen- und Maschinenfabrik A OG (a/k/a FEG)	Budapest, Hungary
jwh	Staaliche Waffenfabrik	Châtellerault, France
K	Luck & Wagner	Suhl, Germany
l	Astrawerk, AG	Chemnitz, Germany
LU	?	?

MAUSER MILITARY RIFLE MARKINGS

Code	Subcontractor	Location
lxr	Dianawerk Mayer & Grammelspacher	Rastatt (Baden), Germany
m	Limbacher Maschinenfabrik, Bach und Winter	Limbach, Germany
n	Elsterwerder Fahrradfabrik E W Reichenbach GmbH	Elsterwerder, Germany
ouj	VDM Luftfahrtwerke A-G, Werk Friedland	Breslau, Germany
s	Dynamit A-G (formerly Alfred Nobel & Co.),	St.-Lambrecht, Germany
S/243	Mauser Werke	Berlin-Borsigwalde, Germany
svw	Mauser Werke	Oberndorf, Germany

MAUSER MILITARY RIFLE MARKINGS

GERMAN POLICE DISTRICT CODES

The following Tables list markings for Bavarian and Prussian police districts according to the regulations of 1930 (for Bavaria) and 1932 (for Prussia). The term "Prussian" can be misleading as, for historical reasons, its geographical extent was much greater than the traditional "Prussian" area around Berlin, extending, in fact, all the way to the borders of Belgium, Luxemburg and the Netherlands in the west and Poland and Czechoslovakia to the east, including the Poland-surrounded entity of East Prussia (*Ostpreussen*). The identification of German states as individual entities (Prussia, Bavaria, and so on) was abolished by decree in January 1934, and was not resumed until after World War II with new administrative areas that, in most instances, are quite different from the pre-War divisions even when the old names are retained.

Marking	Bavarian Police District
A	Augsburg
B	Bamberg
E	Eichstätt
F	Fürstenfeldbruck
H	Hof
K	Kaiserslautern
L	Lindau
Lu	Ludwigshafen
M	München
N	Nürnberg-Fürth
R	Regensburg
Sp	Speyer
W	Würzburg
Z	Zweibrücken

MAUSER MILITARY RIFLE MARKINGS

Marking	Prussian Police District
A.	Aurich
Al.	Allenstein
An.	Aachen
Ar.	Arnsberg
B.	Berlin police administration
Br.	Breslau
D.	Duesseldorf
E.	Erfurt
F.	Frankfurt an der Oder
G.	Gumbinnen
H.	Hannover
Hi.	Hildesheim
K.	Koeslin
Ka.	Kassel
Kg.	Königsberg
Koe.	Köln (i.e., Cologne)
Kz.	Koblenz
Lg.	Lüneburg
Li.	Liegnitz
M.	Muenster
Me.	Merseburg
Mg.	Magdeburg
O.	Osnabrück
Op.	Oppeln
P.	Potsdam
Sd.	Stralsund
Si.	Sigmaringen
St.	Stettin
Sta.	Stade

MAUSER MILITARY RIFLE MARKINGS

Marking	Prussian Police District
T.	Trier
W.	Wiesbaden
Wpr.	Westpreussen

MAUSER MILITARY RIFLE MARKINGS

A MAUSER GAZETTEER

A number of place names appear in this book, as well as in the markings on Mausers themselves. Some of these places are well known, but many others are obscure to varying degrees. This Gazetteer lists places mentioned elsewhere in this book or which are otherwise pertinent to Mausers; a basic description of their locations is given, and here and there some mention is made of a possibly noteworthy fact or a bit of trivia. Places which were or are the site of a factory that either made Mauser rifles or significantly refurbished or reworked them (as opposed to just making rifle parts) are marked with an asterisk.

Altenberg, a city in eastern Germany.
*****Amberg**, a city in southeast Germany, site of the pre-1918 *Königliche bayerische Gewehrfabrik* (Royal Bavarian Rifle Factory).
*****Ankara**, the capital of Turkey.
Bangkok, capital of **Siam** (Thailand). The "Bangkok Era" dating system is used on some Siamese Mausers.
Bayern (English name *Bavaria*), pre-1918 kingdom in southern Germany.
*****Berlin**, the capital of Germany and of Prussia.
Bosnia-Herzegovina, a Balkan republic once part of Yugoslavia; independent since 1992.
Brackwede, a town in western Germany.
*****Brno**, a city in the Czech Republic, site of the ČZ factory.
*****Bruenn** (also spelled **Brünn**), German name for **Brno.**
*****Bystrica** (German name for *Bystřice*), a town in the Czech Republic (formerly Czechoslovakia).
Châtellerault, a French city south of Paris.
Chemnitz, a city in eastern Germany (called *Karl-Marx-Stadt* during the Communist era).

MAUSER MILITARY RIFLE MARKINGS

*Cugir, a city in the Romanian province of Transylvania; site of an arsenal which did a large amount of Mauser refurbishment.

*Danzig, a port city formerly in northeast Germany, a site of the pre-1918 *Königliche Gewehrfabriken* (Royal Arsenals); it is now the Polish city Gdańsk.

Döbeln, a city in eastern Germany.

Dresden, a city in eastern Germany.

Elsterwerda, a town in eastern Germany.

*Erfurt, a city in eastern Germany, a site of the pre-1918 *Königliche Gewehrfabriken* (Royal Arsenals).

*Eskilstuna, a town near Stockholm, Sweden; site of the *Carl Gustafs Gevärsfaktori* (Carl Gustaf Arms Factory).

Freital, a town in eastern Germany.

Hanover, a city in central Germany, site of a *Reichswehr* arms refurbishing facility.

*Hanyang, a city in Hubei Province, central China; now a part of the city of Wuhan..

*Herstal-léz-Liège, a town in Belgium, site of *Fabrique National*.

*Huskvarna, a city in southern Sweden; site of the *Husqvarna Väpenfabrik* (Husqvarna Weapons Factory).

Idar-Oberstein, a town in western Germany.

*Itajubá, a city in Brazil.

*Kırıkkale, a northeast suburb of **Ankara.**

*Kragujevac, a city in Serbia; site of such Yugoslav arsenals as Preduzeće 44, VTZ and Заводи Црвена Застава (*Zavodi Crvena Zastava*, "Red Banner Factories").

Kraków, a city in southern Poland; site of a major arms works.

Kreutzlingen, a town in northeastern Switzerland where

MAUSER MILITARY RIFLE MARKINGS

Mausers with Oberndorf-made components were assembled illegally in 1929-30, in violation of the Versailles Treaty.

***Kwantung**, a region in northeastern China, once a part of **Manchuria**; it is now called Liaoning. The capital of Liaoning is Shenyang, formerly called **Mukden**.

***La Coruña**, a city in northern Spain.

***Liège**, a Belgian city, home of FN (at **Herstal-léz-Liège**).

  Limbach, (in this case **Limbach-Oberfrohna**) a town in eastern Germany.

***Lüttich**, German name for **Liège**.

***Manchuria**, a region in northeastern China, site of the **Mukden** arsenal.

***Mexico City, D.F.** (Spanish name *Ciudad de México, Districto Federal*), capital of Mexico and site of the National Arms Factory (*Fábrica Nacional de Armas*).

  Minden, a city in western Germany.

***Mukden**, a city in **Manchuria**; it is now called Shenyang, and is the capital of Liaoning province.

***Niederschönweide bei Berlin**, home of Mauser maker *Waffenwerk Oberspree, Kornbusch & Co.*

  Nürnberg (English name *Nuremberg*), a city in southern Germany.

***Oberndorf am Neckar**, a town on the River Neckar in southern Germany, site of the Mauser brothers' factory.

  Oberschönau, a city in eastern Germany.

***Otterup**, a town on the island of Fyn in Denmark; ex-German Mausers were refurbished and altered here after World War II.

***Oviedo**, a city in northern Spain.

***Paris**, the capital of France; its St.-Denis district was the site of *Société Française des Armes Portatives* (French Small Arms Co.), which produced the

MAUSER MILITARY RIFLE MARKINGS

Daudetau-Mauser conversion for Uruguay at the end of the 19th century.

*Persia, a country in the Middle East; it changed its name to Iran in 1935.

Prussia (German name *Preussen*), a pre-1918 kingdom; the capital was Berlin.

*Radom, a city in southern Poland; site of an important arms factory complex.

*Ramat ha-Sharon, early arms factory site in Israel; K98k Mausers were made here in the 1950s. The factory is now owned by IMI (Israel Military Industries).

Rastatt, a city in southwest Germany.

Remscheid, a city in western Germany.

*Rosario, a city in Argentina; one of two sites of domestic Argentine Mauser production (the other was at **Santa Fe**).

[the] Ruhr, the most important industrial region in western Germany.

Sachsen (English name *Saxony*), a pre-1918 kingdom in eastern Germany.

*Santa Fe, a city in Argentina, site of one of two domestic Argentine Mauser factories.

Siam, a country in southeast Asia; it changed its name to Thailand (*Prathet Thai*) in 1938.

*Spandau, a district of Berlin, a site of site of the pre-1918 *Königliche Gewehrfabriken* (Royal Arsenals)

*Steyr, a city in Austria, home of the Mauser makers *Österreichische Waffenfabrik-Gesellschaft* (OE WG) and Steyr-Daimler-Puch, A-G.

*Suhl, a city in eastern Germany, home of Mauser makers Simson & Sohn, C. G. Haenel, and V. Ch. Schilling.

*Tarrasa, a city in eastern Spain; site of *Industrias de Guerra de Cataluña* (Catalonian War Industries).

MAUSER MILITARY RIFLE MARKINGS

Thüringen (English name *Thuringia*), a region in eastern Germany.

*****Tokyo**, capital of Japan. Some Siamese Mausers were made here under license in the early 20th century, possibly at the Koishikawa arsenal.

*****Warszawa** (Polish name for *Warsaw*), the capital of Poland.

Witten, a city in western Germany.

Württemberg (also spelled **Wuerttemberg**), a pre-1918 kingdom in southwest Germany; now part of the state of Baden-Württemberg. **Oberndorf am Neckar** is here.

Wuppertal, a city in western Germany.

Zaandam, city in the Netherlands.

Zeithain, a place in Germany, site of a Reichswehr arms refurbishing facility.

MAUSER MILITARY RIFLE MARKINGS

SELECTED BIBLIOGRAPHY

Allan, Francis C., and Roger L. Wakelam, *The Siamese Mauser; A Study of Siamese/Thai Type 46 Rifles & Type 47 Carbines*, Palm Coast, FL:Francis C. Allan, 1987

Ball, Robert W.D., *Mauser Military Rifles of the World*, 2nd Ed., Iola, WI:Krause Publications, 2000

Bester, Ron, *Boer rifles and carbines of the Anglo-Boer War*, Bloemfontein, South Africa:War Museum of the Boer Republics, 1994

Goertz, J. and D. Bryans, *German Small Arms Markings From Authentic Sources*, Marceline, MO:Walsworth Publishing Co., 1997

Gwóźdź, Zbigniew, and Piotr Zarzycki, *Polskie konstrukcje broni strzeleckiej*, Warsaw:SIGMA NOT, 1993

Hoffman, Richard and Noel Schott, *Handbook of Military Rifle Marks, 1866-1950*, 2nd Ed., St. Louis, MO: Mapleleaf Militaria Publication, 1998

Kehaya, Steve, and Joe Poyer, *The Swedish Mauser*, Tustin, CA:North Cape Publications, 2000

Królikiewicz, Tadeusz, *Bagnety*, Warsaw:Bellona, 1998

Law, Richard D., *Backbone of the Wehrmacht, The German K98k Rifle, 1934-1945*, 1993 Author's Revised Edition, Cobourg, Ont., Canada:Collector Grade Publications, 1998

MAUSER MILITARY RIFLE MARKINGS

Law, Richard D., *Backbone of the Wehrmacht, Vol. II; Sniper Variations of the German K98k Rifle*, Cobourg, Ont., Canada:Collector Grade Publications, 1996

Moudrý, Petr, *Bodáky Československa*, Prague:ARS-ARM, 1992

Olson, Ludwig, *Mauser Bolt Rifles,* 3rd Ed., Montezuma, IA:F. Brownell & Son, Publishers, 1976

Pere, Nuri, *Osmalılarda Madenî Paralar*, İstanbul:Yapı ve Kredi Bankası, 1968

Walter, John, *Rifles of the World*, Northbrook, IL:DBI Books, 1993

Zielinski, Stan and Bin Shih, "Chinese Model 77 Rifle", *The Military Rifle Journal*, Issue 118, October 2000

MAUSER MILITARY RIFLE MARKINGS

NOTES